Applications of Group Theory in Cryptography

Post-quantum Group-based Cryptography

Mathematical
Surveys
and
Monographs

Volume 278

Applications of Group Theory in Cryptography

Post-quantum Group-based Cryptography

Delaram Kahrobaei
Ramón Flores
Marialaura Noce
Maggie E. Habeeb
Christopher Battarbee

AMS AMERICAN MATHEMATICAL SOCIETY
Providence, Rhode Island

2020 *Mathematics Subject Classification.* Primary 94A60, 81P94, 20F05, 20F99, 20F10.

For additional information and updates on this book, visit
www.ams.org/bookpages/surv-278

Library of Congress Cataloging-in-Publication Data

Names: Kahrobaei, Delaram, 1975- author. | Flores, Ramón, 1975- author. | Noce, Marialaura, 1992- author. | Habeeb, Maggie E., 1983- author. | Battarbee, Christopher, 1998- author.

Title: Applications of group theory in cryptography : post-quantum group-based cryptography / Delaram Kahrobaei, Ramón Flores, Marialaura Noce, Maggie E. Habeeb, Christopher Battarbee.

Description: Providence, Rhode Island : American Mathematical Society, [2024] | Series: Mathematical surveys and monographs, 0076-5376 ; volume 278 | Includes bibliographical references and index.

Identifiers: LCCN 2023044735 | ISBN 9781470474690 (paperback) | ISBN 9781470476212 (ebook)

Subjects: LCSH: Group theory. | Cryptography. | AMS: Information and communication, circuits – Communication, information – Cryptography. | Quantum theory – Axiomatics, foundations, philosophy – Quantum cryptography. | Group theory and generalizations – Special aspects of infinite or finite groups – Generators, relations, and presentations. | Group theory and generalizations – Special aspects of infinite or finite groups – None of the above, but in this section. | Group theory and generalizations – Special aspects of infinite or finite groups – Word problems, other decision problems, connections with logic and automata.

Classification: LCC QA174.2 .K34 2024 | DDC 652/.8015122–dc23/eng20231103

LC record available at https://lccn.loc.gov/2023044735

∞ The paper used in this book is acid-free and falls within the guidelines
established to ensure permanence and durability.
Visit the AMS home page at `https://www.ams.org/`

10 9 8 7 6 5 4 3 2 1 29 28 27 26 25 24

To the memories of Gilbert Baumslag and Benjamin Fine
and to Michael Anshel

Contents

Introduction

Group-based cryptography is a very recent and fast-growing field of research at the intersection of group theory, combinatorics, complexity theory, coding theory, and cryptology. The main goal of group-based cryptography is the development and analysis of secure cryptographic protocols using as input algorithmic properties of different classes of groups, and in particular the complexity of decision and search problems in finitely generated groups. In general these techniques are especially useful in public-key cryptography, which provides a method for two parties to communicate privately, despite the lack of any prearranged security keys.

The methodology of group-based cryptography follows the usual pattern of showing that a *break* to a cryptosystem amounts to the solution of a difficult mathematical problem. Indeed, one can think of the very first examples of public-key cryptography as *group-based* — after all, the famous problems of integer factorization or recovering discrete logarithms can be thought of as group-theoretic problems, and cryptosystems based on these problems retain a vital role in modern digital infrastructure. However, with the advent of quantum computers, these underlying mathematical problems are efficiently solvable, and hence the corresponding cryptographic schemes are deemed insecure in this setting. In the post-quantum era different tools are required: in this direction, in 2015, the National Security Agency (NSA) and the National Institute of Standards and Technology (NIST) announced an initiative for the development of post-quantum cryptosystems. Lattice-based, code-based, multivariate-based, and hash-based primitives where shortlisted as third-round finalists suggested for standardization five years later [**170**]; nevertheless, since the announcement some of these finalists have been the subject of major cryptanalytic work — in some cases casting the viability of these schemes as post-quantum standards into serious doubt. In light of these developments it remains vital to study a diverse range of computational problems; a promising class of such tools comes from group-based cryptography. In fact, not only does this theory offer a rich supply of complex and varied problems for cryptography; reciprocally, the study of cryptographic algorithms built from these problems has contributed results to computational group theory.

In the following we provide a nonexhaustive but very significant list of families of groups which have been considered as platforms for different security protocols: braid groups ([**2**] and [**131**]), polycyclic groups [**46**], matrix groups [**90**], nonabelian simple groups [**174**], hyperbolic groups [**30**], arithmetic groups [**118**], graph groups [**61**], nilpotent groups [**122**], automata groups ([**71**], [**86**]), Thompson groups [**127**], free metabelian groups [**197**], small cancellation groups [**97**], free nilpotent p-groups [**120**], Engel groups [**119**], or infinite pro-p groups [**121**]. Many of these families are discussed in the last part of the book.

Computational problems thought to be difficult in these groups give rise, in turn, to a number of cryptographic protocols. Among them, we quote here the group-based digital signature WalnutDSA^{TM} [**3**], which has faced several attacks ([**99**], [**13**], [**136**]) and is claimed by the authors to be quantum-resistant; the AAG key exchange protocol [**2**]; a noncommutative version of Diffie-Hellman [**131**]; homomorphic encryption schemes ([**90**], [**174**]), hash functions [**177**], multilinear maps [**122**], noncommutative secret-sharing schemes ([**175**], [**97**]), authentication schemes [**60**], or asymmetric key exchange protocols [**30**]. These schemes, and their corresponding cryptanalysis, are the subject of Part II of the book. We also provide a brief overview of quantum cryptanalysis for group-based cryptography, including Shor's algorithm [**194**] and Kuperberg's algorithm for the dihedral hidden subgroup problem [**139**]. Moreover, some of the techniques are used to establish a strong connection to the area of group-based cryptography established in [**36**].

Our book is intended to be a comprehensive introduction to group-based cryptography. We note that there already exist thorough references on the subject, such as the monographs of Myasnikov-Shpilrain-Ushakov [**160**] and [**162**]. Our treatment is not so exhaustive, as we aim to offer a friendly approach which is accessible to both computer scientists and mathematicians. With this goal in mind, we have tried to present the group-theoretic statements and proofs in the easiest way, avoiding unnecessary technicalities and preferring, in general, the combinatorial versions of the results; in turn, we have been very descriptive when developing the protocols, the algorithms, and the cryptanalysis. On the other hand, our intention has also been to present the most updated version of the subject, with special attention to the protocols and references of the last years. If we have achieved the goal of adding a new item to the bibliography where the material is accessible and completely updated, we will be satisfied as authors.

We conclude this introduction by providing a detailed description of the contents of our book.

Organization of the book. The book is divided into three parts. The first one provides background information and basic notions of group theory, algorithmic problems in groups, and classical cryptography. More precisely, in Chapter 1, we survey the basic theory of groups, subgroups, and special classes of groups such as cyclic and nilpotent groups. In this part, we also describe free groups and combinatorial group theory. Chapter 2 lays the groundwork for describing algorithmic problems in group theory: we describe group-theoretic decision problems including the word, conjugacy, and isomorphism decision problems. We also describe the power decision problem, the endomorphism decision problem, the n-th root decision problem, and the geodesic length decision problem. We also present many search problems. In Chapter 3 we describe a number of cryptosystems that have been built around several classes of groups in the last decades. In particular, we present digital signature schemes, secret sharing schemes, and multilinear maps.

Part 2 is devoted to group-theoretic post-quantum cryptography. In Chapter 4 we present several key exchange protocols based on an underlying noncommutative structure such as Anshel-Anshel-Goldfeld (AAG) key exchange, Ko, Lee, et al. key exchange, and the noncommutative ElGamal key exchange. We also introduce two key exchange protocols based on the semidirect product of groups or semigroups. A discussion of noncommutative digital signatures, such as the Kahrobaei-Koupparis digital signature scheme and the Mohan and Anjaneyulu digital signature scheme, and of noncommutative secret sharing schemes is also included. We discuss the Panagopoulos threshold scheme as it has motivated the construction of some noncommutative threshold schemes. These include the Habeeb-Kahrobaei-Shpilrain (HKS) secret sharing scheme, which is a hybrid of Shamir's secret sharing scheme that provides some advantages to Shamir's scheme, and threshold schemes based on Nielsen transformations. Chapter 5 introduces some recent attacks against these schemes: for example the length-based attack, the linear centralizer attack, the linear decomposition attack, and the quotient attack. Chapter 6 deals with quantum cryptanalysis of some examples of group-based cryptography, in particular as it relates to complexity of the hidden subgroup problem (HSP). We cover the famous construct of Shor's algorithm and survey the state of the art for quantum algorithms addressing the hidden subgroup problem in groups not accounted for by Shor's algorithm. Finally, we provide a "case study"-type exposition of two recent results in group-based quantum cryptanalysis.

The last part of the book is devoted to suitable platforms for the cryptographic protocols presented in Part 2. In each chapter we give the definition of the class of groups proposed together with important properties and the status of their algorithmic problems. Then, we survey the current status of the respective group in cryptography, and we also include a list of open problems, which we hope will guide researchers who wish to work in this field. The groups that we propose as cryptographic platforms are the following: braid groups, hyperbolic groups, small cancellation groups, polycyclic groups, graph groups, arithmetic groups, Engel groups, and self-similar groups.

Acknowledgments

Delaram Kahrobaei: Over the years, this research has been supported, in various forms, by grants from the National Science Foundation (NSF), the Office of Naval Research, Canada's New Frontiers in Research Fund (NFRF)-Exploration stream, the American Association for the Advancement of Science (AAAS), and the London Mathematical Society (LMS). The first author is grateful to the Institut des Hautes Études Scientifiques (IHES) for providing excellent scientific environment to do mathematics, this article was finished during her visit to IHES.

Ramón Flores: We warmly thank Y. Antolín and J. González-Meneses for helpful comments. The second author was partially supported by grant PID2020-117971GB-C21 from the Spanish Ministry for Science and Innovation (MCIN), and grant FQM-213 from the Junta de Andalucía.

Notation

Throughout the book we will use standard notation and terminology (see [**186**]). We often consider nonabelian groups with multiplicative notation, in which the multiplication symbol is always omitted. In the following table G always denotes a group.

Groups	
Trivial element of G	e, 1, 1_G
G is isomorphic to K	$G \simeq K$
H is a subgroup of G	$H \leq G$
Subgroup generated by X in G	$\langle X \rangle$
Normal closure of X in G	$\langle X \rangle^G$
The index of a subgroup H in G	$\lvert G : H \rvert$
g and h are conjugate elements	$g^h = h^{-1}gh$
Commutator of elements h and g	$[h, g] = h^{-1}g^{-1}hg$
Commutator of two sets A and B	$[A, B]$
Commutator (or derived) subgroup of G	$G' = [G, G]$
Center of G	$Z(G)$
Centralizer of a set X in G	$C_G(X)$
Normalizer of a set X in G	$N_G(X)$

Part 1

Background Information

CHAPTER 1

Group Theory

1.1. Introduction

In this introductory chapter we state the main group-theoretic definitions and results that will be used extensively throughout the book. This chapter is geared towards nonmathematicians; thus, it begins with elementary definitions. For the interested reader, we refer to the text [**186**] for a thorough introduction of the topic.

1.2. Basic definitions and examples

We will begin by stating the definitions of a group and a group homomorphism.

DEFINITION 1.1. *A* group *is a nonempty set G endowed with a binary operation $*: G \times G \to G$ such that the following axioms hold:*

(a) ***Associativity:*** *for all elements $x, y, z \in G$ we have*

$$(x * y) * z = x * (y * z).$$

(b) ***Identity:*** *for all elements $x \in G$ there exists a unique element $e \in G$, called the* identity *of G, such that*

$$e * x = x * e.$$

(c) ***Inverses:*** *for all elements $x \in G$ there exists a unique element $x^{-1} \in G$ such that*

$$x * x^{-1} = x^{-1} * x = e.$$

The element x^{-1} is called the inverse *of x.*

Additionally, if for a group G the following axiom holds:

(d) ***Commutativity:*** *for every $x, y \in G$, we have*

$$x * y = y * x.$$

We say that the group G is commutative *or* abelian.

When the operation of a group needs to be emphasized it is written $(G, *)$ rather than G.

REMARK 1.2. It is customary to omit $*$ and write just xy, or $x+y$ if the group is abelian. The identity element is usually written $e = 1$ in the general case, and $e = 0$ in the abelian case.

By removing some of the group axioms, different algebraic structures arise. As they will be mentioned later in the text, we define them here.

DEFINITION 1.3. *A* semigroup *is a nonempty set G endowed with an associative binary operation, $*: G \times G \to G$.*

DEFINITION 1.4. *A* monoid *is a semigroup with an identity element.*

Maps between groups that preserve the group structure are called *homomorphisms*. Formally, a homomorphism is defined as follows.

DEFINITION 1.5. *Let* $(G, *)$ *and* $(H, \circ)$ *be two groups. We say that a map* $f : G \to H$ *is a* homomorphism of groups *if for every* $x, y \in G$, $f(x*y) = f(x)\circ f(y)$.

Injective homomorphisms are called *monomorphisms*, while surjective homomorphisms are called *epimorphisms*. Moreover, bijective homomorphisms are called *isomorphisms*. When a map is an isomorphism, it is guaranteed that the inverse map is also an isomorphism. When there is an isomorphism between two groups G and G', the groups are called isomorphic and denoted $G \cong G'$. The notion of isomorphic groups is the analog of equality in the algebraic setting. It should be noted that an isomorphism between groups need not be unique; there can be many different isomorphisms between two groups. Note that every group is isomorphic to itself (take, for example, the identity map). An isomorphism from a group G to itself is called an *automorphism*, while a homomorphism from a group G to itself is called an *endomorphism*.

1.3. Examples of groups

In this section we will introduce several elementary examples of groups to illustrate the notions presented in the previous section.

1.3.1. Integer, rational, and real numbers. The set of integers under addition $(\mathbb{Z}, +)$, the set of rational numbers under addition $(\mathbb{Q}, +)$, and the set of real numbers under addition $(\mathbb{R}, +)$ are abelian groups. Notice that none of these form a group under multiplication. However, removing 0 from the set of rational numbers $\mathbb{Q}$ and the set of real numbers $\mathbb{R}$ results in groups under multiplication.

1.3.2. Cyclic groups. A group G is *cyclic* if there exists an element x such that for every $y \in G$ there exists $n \in \mathbb{Z}$ such that $x^n = y$. The element x is called a *generator* of G. It is not hard to see that if G is cyclic infinite, it is isomorphic to the group $\mathbb{Z}$ of the integer numbers. On the other hand, if G is cyclic finite, it is isomorphic to the group $\mathbb{Z}_n$ of integers modulo n for some $n \geq 1$. If an element $g \in \mathbb{Z}_n$ is a generator of the group, then g is called a primitive root modulo n. Clearly, every cyclic group is abelian.

1.3.3. Direct product. Given two groups G and H, the Cartesian product $G \times H$ can be endowed with a natural group structure given by $(a, b)(c, d) = (ac, bd)$. This is called the *direct product* of G and H.

1.3.4. Symmetric groups. Consider a set X of n elements. Then the set of permutations of the elements of X under the operation composition defines a group called the *symmetric group*. Symmetric groups are usually denoted by Σ_n or S_n. Cayley's theorem (see **[186]** for more details) shows that every finite group is isomorphic to a subgroup of S_n.

1.3.5. Linear groups. In order to define linear groups, we begin with a definition.

DEFINITION 1.6. *A field is a set* $\mathbb{F}$ *together with two operations, addition and multiplication, such that*

- $(\mathbb{F}, +)$ *is an abelian group with identity element* 0.

- $(\mathbb{F} - \{0\}, \cdot)$ *is an abelian group.*
- *For all* $x, y, z \in \mathbb{F}, (x + y) \cdot z = x \cdot z + y \cdot z$ *and* $z \cdot (x + y) = z\dot{x} + z \cdot y$.

Classical examples of fields are $\mathbb{Q}$, $\mathbb{R}$, or $\mathbb{Z}_p$, for p prime.

Let $n \geq 2$. Consider the set of all $n \times n$ matrices with entries in the field, $\mathbb{F}$. One can easily see that this set is associative under multiplication. Denote the matrix that has ones in the diagonal and zeros elsewhere by I. It is clear that for all matrices A, $AI = IA = A$. Hence, I is the multiplicative identity. If we restrict the set to $n \times n$ matrices over the field $\mathbb{F}$ which have inverses, namely those whose determinant is nonzero, then this set forms a group under multiplication. This group is called *the general linear group of dimension* n over $\mathbb{F}$, and is denoted by $GL(n, \mathbb{F})$. Note that in general this group is nonabelian.

Consider the matrices from $GL(n, \mathbb{F})$ with determinant 1. Using properties of determinants, it can be shown that this set is also a group under multiplication. This group is called the *special linear group*, and is denoted $SL(n, \mathbb{F})$.

Throughout the book, many other families of groups will be introduced together with their importance in cryptography. They will be defined in due time.

1.4. Subgroups and order of elements

Next we define the natural subobject in the context of groups: a subgroup.

DEFINITION 1.7. *Let* $(G, \cdot)$ *be a group and* H *a nonempty subset of* G. *If* H *is a group relative to the binary operation of* G, *then it is said that* H *is a subgroup of* G *and we denote it by* $H \leq G$ *(*$H < G$ *if the inclusion is strict).*

EXAMPLE 1.8. *Here we give examples of subgroups based on the groups of the previous section.*

- *The groups* G *and* $\{e\}$ *are trivially subgroups of any group* G.
- *The set of integers* $\mathbb{Z}$ *is a subgroup of the set of rational numbers,* $\mathbb{Q}$, *under addition. Moreover,* $\mathbb{Q}$ *is a subgroup of* $\mathbb{R}$ *under addition as well as multiplication.*
- *For every* n, S_n *is a subgroup of* S_{n+1} *and the inclusion is strict; that is,* $S_n < S_{n+1}$.
- *For every* $n \geq 2$ *and every field* $\mathbb{F}$, *we have* $SL(n, \mathbb{F}) \leq GL(n, \mathbb{F})$.

For simplicity, the following proposition can be used to show that a subset of a group is a subgroup. The proof is immediate using the definition of subgroup.

PROPOSITION 1.9. If H is a nonempty subset of the group G, then the following three conditions are equivalent:

(a) $H \leq G$.
(b) For all elements x, y is in H, $xy \in H$ and $x^{-1} \in H$.
(c) For all $x, y \in H$ we have $x^{-1}y \in H$.

As a group is in particular a set, the cardinality of the set is an important property.

DEFINITION 1.10. *Let* G *be a group. If* G *is finite, its cardinality (as a set) is called the* order *of* G, *and denoted* $|G|$ *or* $o(G)$; *if it is not finite, we will say that* $|G| = \infty$. *Moreover, given an element* x *in a group* G, *the order* $|x|$ *of* x *is the smallest* n *such that* $x^n = e$. *If such an* n *does not exist, we will say that* $|x| = \infty$.

For finite groups, the order of the elements in the group and the order of the subgroups must satisfy the requirements indicated in the following theorem and corollary.

THEOREM 1.11 (Lagrange). *The order of any subgroup of a finite group divides the order of the group.*

COROLLARY 1.12. *Let G be a finite group of order n. Then for every element $x \in G$ we have $x^n = e$.*

Now that we have stated the basic definitions and properties about the order of elements in a group, we would like to introduce some terminology. Given a group G and an integer $m > 1$, if we have $x^m = e$ for every $x \in G$, we say that G is *m-torsion* or simply *torsion* or *periodic*. If, on the other hand, every element except the identity has infinite order, we say that G is *torsion free*.

To conclude this section, we will introduce a way to present a group without explicitly listing its elements. This notion will be very important throughout this book. We begin by stating a necessary result.

PROPOSITION 1.13. *Given a group G, the intersection of any collection of subgroups of G is again a subgroup.*

This proposition gives rise to the following definition.

DEFINITION 1.14. *Let X be a subset of a group G. Then the intersection of all the subgroups that contain X is called the subgroup* generated *by X, which is denoted $\langle X \rangle$. The elements of X are called the* generators *of $\langle X \rangle$.*

The following result gives a practical way of calculating the elements of a subgroup given a generating set X.

THEOREM 1.15. *Let X be a subset of a group G. Then the subgroup $\langle X \rangle$ generated by X consists of all possible products of elements in X and their inverses.*

A subgroup G is called *finitely generated* if it possesses a finite generating set X. Many of the groups that will be presented in this book are finitely generated.

1.5. Normal subgroups and quotients

In this section we will define the important notion of *quotient group* (or *factor group*). In order to define a quotient groups we must first introduce three concepts: cosets, index of a subgroup, and conjugation.

DEFINITION 1.16. *Let H be a subgroup of a group G. Given an element $a \in G$, we define the* left coset *of H in G determined by a as*

$$aH = \{ah \mid h \in H\}.$$

In turn, the corresponding right coset *of H in G determined by a is $Ha = \{ha \mid h \in H\}$.*

DEFINITION 1.17. *For a subgroup $H \leq G$, the cardinality of the set of left cosets of H in G is called the* index *of H in G, and is usually denoted $[G : H]$ or $|G : H|$.*

We will frequently make use of the crucial notion of conjugation, which is defined as follows.

DEFINITION 1.18. *Given two elements g and h in a group G, we define the* conjugate *of g by h as the element $g^h = hgh^{-1}$.*

Using this notion, every element h of G defines an automorphism c_h of G given by $c_h(g) = g^h$. To see this not that this map clearly preserves the group structure and has as its inverse $c_{h^{-1}}$.

REMARK 1.19. It is also common in the literature to denote $g^h = h^{-1}gh$. In this book, if it is not explicitly stated, we will use the notation of Definition 1.18.

Now that the necessary definitions have been stated, we can define normal subgroups.

DEFINITION 1.20. *Let N be a subgroup of a group G. If for every $g \in G$ and $n \in N$ we have $g^n \in N$, then N is said to be* normal *in G, and we denote $N \trianglelefteq G$.*

Equivalently, N is a normal subgroup of G if and only if any right coset of N in G is also a left coset of N in G, that is, $xN = Nx$ for every $x \in G$.

EXAMPLE 1.21. *Here, we provide some standard examples of normal subgroups.*

- *If G is any group, then the trivial subgroup $\{e\}$ and G itself are normal in G.*
- *If G is an abelian group, then, as $xH = Hx$ for any $x \in G$, every subgroup H of G is normal in G. Note that the converse is not necessarily true.*
- *The special linear group is a normal subgroup of the general linear group. This is a consequence of the properties of the product of determinants.*

Another important example of a normal subgroup that occurs in every group is the center of a group, which is defined below.

DEFINITION 1.22. *Let G be a group. Then the abelian subgroup*

$$\{x \in G \mid xy = yx \text{ for every } y \in G\}$$

is called the center *of the group G, and is denoted $Z(G)$.*

The following two notions, derived from previous definitions, will be also important throughout the book.

DEFINITION 1.23. *Let G be a group and let X be a nonempty subset of G. Then the* normalizer, *denoted $N_G(X)$, of X is the set of elements $g \in G$ such that x^g belongs to X for every $x \in X$. In turn, the* centralizer, *denoted $C_G(X)$, is the set of all $g \in G$ such that $xg = gx$ for every $x \in X$. That is, the centralizer is the set of elements of G that commute with all the elements of X.*

Observe that for every $X \subseteq G$, $N_G(X)$ and $C_G(X)$ are subgroups of G. Moreover, $C_G(X) \leq N_G(X)$. Frequently, the normalizer and centralizer of a subset X are considered when X is itself a subgroup of G. Note that unlike the center of G (which is the centralizer of G, in fact), centralizers are not always abelian.

We finish this section studying the important concept of a quotient group. Let N be a normal subgroup of a group G. We denote the set of all cosets of N in G, $\{gN : g \in G\}$, by G/N. Then the set G/N with the product defined by $g_1Ng_2N = g_1g_2N$ forms a group.

DEFINITION 1.24. *$G/N = \{gN : g \in G\}$ together with the operation defined by $g_1Ng_2N = g_1g_2N$ is called quotient group of G by N.*

The quotient operation provides a useful way to construct new groups out of normal subgroups of a given group.

1.6. Group actions

In this section we will define group actions, which will arise in various portions of the text. For the interested reader, [**204**] provides a thorough explanation of group actions.

DEFINITION 1.25. *Let G be a group and X a set. Then a* group action *of G over X (on the left) is a map $G \times X \to X$ that takes every pair (g, x) to an element gx of X, in such a way that:*

- *For every $g, h \in G$ and $x \in X$ we have $(gh)x = g(hx)$.*
- *$ex = x$ for every $x \in X$ where e is the identity element of G.*

Actions on the right are defined in an analogous way.

This definition is equivalent to to the following. Let X be a set, $Aut(X)$ the group of bijections of X. Then an action of G over X is simply a group homomorphism $G \to Aut(X)$. We would like to note that if $|X| = n$, then $Aut(X)$ is the symmetric group S_n. Additionally, if the set X carries some structure, it is common to require some additional hypotheses over the map $G \times X \to X$. For example, if X is a topological space, the map will be continuous; if it is a differentiable manifold, the map will be differentiable, etc.

EXAMPLE 1.26. *Below we provide some examples of group actions.*

- *The group $\mathbb{Z}$ acts on the real line by translations. In turn, $\mathbb{Z}_2$ acts on the real line by symmetry with respect to the origin.*
- *The symmetric group S_n acts on any set of n elements by permutations.*
- *The group $\mathbb{Z}_n$ acts on the regular polygon of n sides by a rotation of $2\pi/n$ radians.*
- *The general linear group $GL(n, \mathbb{R})$ acts on the vector space $\mathbb{R}^n$ by matrix multiplication.*

We are now ready to define the notions of an orbit, a stabilizer, and fixed points of a group actions, which are important this the context.

DEFINITION 1.27. *Let G be a group acting over a set X.*

(a) *Then given an element $x \in X$, the* orbit *of x is the set of elements gx, with $g \in G$.*
(b) *The* stabilizer *G_x of x (also called the* isotropy group*) is the subgroup of G formed for the elements $g \in G$ such that $gx = x$.*
(c) *A point $x \in X$ such that $gx = x$ for every $g \in G$ is called a* fixed point *of the action.*

We will conclude this section by introducing terminology for certain properties of a group action.

DEFINITION 1.28. *Let G be a group acting over a set X. Then the action is said to be:*

- Effective, *if for every nontrivial $g \in G$ there exists $x \in X$ such that $gx \neq x$.*
- Free, *if for every $x \in X$ and every nontrivial $g \in G$, $gx \neq x$.*
- Transitive, *if for every $x, y \in X$ there exists $g \in G$ such that $gx = y$.*

1.7. Extensions

An extension of a group is a common way of describing a group G by using two "simpler" groups. In the case when G is finite and the extension is nontrivial, these "simpler" groups have smaller cardinality. There will be several particular instances of extensions that are important throughout the book.

DEFINITION 1.29. *Let $f : G \to H$ be a group homomorphism.*

(a) *The set $\{x \in G : f(x) = e\}$ is a subgroup G called the* kernel *of f, and is denoted $Ker\ f$.*
(b) *Let H be a subgroup of G. The set $\{x \in H : f(y) = x$ for some $y \in G\}$ is a subgroup of G called the the* image *of f, and is denoted $Im\ f$.*

It is straightforward to check that $Ker\ f$ is a normal subgroup of G.

We are now ready to formally define an extension.

DEFINITION 1.30. *Given two groups K and H, an* extension *of K by H is a group G such that there exists a sequence of two homomorphisms*

$$K \xrightarrow{f} G \xrightarrow{g} H$$

such that:

- $K = Ker\ f$.
- $Im\ f = Ker\ g$.
- $G = Im\ g$.

Sometimes this is referred to as a *short exact sequence* of groups. Observe that given a group G and a normal subgroup K of G, the sequence $K \to G \to G/K$ presents G as an extension of K by G/K.

EXAMPLE 1.31. *Here we provide some examples of extensions.*

- *For every group G, the sequence $G \xrightarrow{f} G \times G \xrightarrow{g} G$, where f is the inclusion of the first factor and g the projection on the second, is an extension.*
- *Given $n > 1$, let $n\mathbb{Z}$ be the set of multiples of n in $\mathbb{Z}$, which is a subgroup of $\mathbb{Z}$. Let $\mathbb{Z}_n$ be the group of integers under addition modulo n. Then $n\mathbb{Z} \to \mathbb{Z} \to \mathbb{Z}_n$ is an extension. Also, observe that $n\mathbb{Z}$ is isomorphic to $\mathbb{Z}$.*

The following type of extension will be very useful through this book as it is used in various cryptographic protocols.

DEFINITION 1.32. *Let K and H be groups, and suppose that H acts on K. Then we define the following operation in the direct product $K \times H$ (considered as sets):*

$$(a, g) * (b, h) = (a(gb), gh).$$

With this action $K \times H$ is a group, and is called the semidirect product $K \rtimes H$ *(with respect to the action).*

It is clear that $K \to K \rtimes H \to H$ is an extension of groups when the first homomorphism is given by inclusion of the first factor and the second by projection of the second factor.

EXAMPLE 1.33. *Suppose the conditions of the previous definition are satisfied. Let n be a natural number with $n \geq 2$. Let H be a group that acts on the set X_n of n elements. If G is any other group and K is the direct product of n copies of G, the action of H over X_n gives rise to an action over K by interchanging of copies of G. The semidirect product of K by H with respect to this action is called the* wreath product *of K by H, and denoted $K\ wr\ H$.*

Note that in this scenario H can be any subgroup of the symmetric group S_n.

1.8. Nilpotent groups

In this section, we recall the definition and the basic notions concerning nilpotent groups. Roughly speaking, the notion of nilpotency generalizes commutativity.

Let G be a group and $x, y \in G$. We define the *commutator* of x and y as the product

$$[x, y] = x^{-1}y^{-1}xy = (y^{-1})^x y.$$

More generally, a commutator of weight $n \geq 1$ of elements $x_1, \ldots, x_n \in G$ is defined recursively as follows:

$$[x_1, \ldots, x_n] = x_1 \text{ if } n = 1, \qquad [x_1, \ldots, x_{n-1}, x_n] = [[x_1, \ldots, x_{n-1}], x_n] \text{ if } n > 1.$$

It is not difficult to check that the following relations hold.

LEMMA 1.34. *Let G be a group and $x, y, z \in G$. Then:*
(a) $[y, x] = [x, y]^{-1}$;
(b) $[x^{-1}, y] = [y, x]^{x^{-1}} = [x, y]^{-x^{-1}}$ *and* $[x, y^{-1}] = [y, x]^{y^{-1}} = [x, y]^{-y^{-1}}$;
(c) $[xy, z] = [x, z][x, z, y][y, z]$ *and* $[x, yz] = [x, z][x, y][x, y, z]$.

Before defining a nilpotent group, we must introduce a subgroup that is necessary in the theory of nilpotent groups. Let H, K be subgroups of G. We define $[H, K] = \langle [h, k] \mid h \in H, k \in K \rangle$ to be the subgroup of G generated by all commutators of the form $[h, k]$.

DEFINITION 1.35. *Consider the following series of subgroups $\gamma_i(G)$, $i = 0, 1, \cdots$.*

$$\gamma_0(G) = G, \gamma_{i+1}(G) = [\gamma_i(G), G]$$

where $\gamma_{i+1}(G) \leq \gamma_i(G)$. A group G is nilpotent of class m if m is the least integer such that $\gamma_m(G) = 1$.

An equivalent way to define nilpotent groups is given by the following result.

PROPOSITION 1.36. *Let G be a group. We say that G is* nilpotent *if and only if G possesses a central series. In other words, G is nilpotent if and only if there exists a chain of subgroups $H_1, \ldots, H_n$ of G,*

$$1 = H_0 \trianglelefteq H_1 \trianglelefteq \cdots \trianglelefteq H_n = G,$$

such that for any $i \in \{0, \ldots, n\}$, H_i normal in G and $H_{i+1}/H_i \leq Z(G/H_i)$, where $Z(G/H_i)$ is the center of G/H_i.

If a group G is nilpotent, the minimal length of its central series is said to be the *nilpotency class* of G and it is denoted by $\mathrm{cl}(G)$. Note that nilpotent groups of class 1 are abelian groups. Moreover, it is important to note that every subgroup and every quotient group of a nilpotent group is nilpotent. Furthermore, the direct product of a finite number of nilpotent groups is nilpotent.

EXAMPLE 1.37. *Here, we present a well-known example of a nilpotent group of class 2: the* **Heisenberg group**. *Let $r, s, t \in \mathbb{Z}$, now consider the group H of the matrices of the form*

$$\begin{pmatrix} 1 & r & s \\ 0 & 1 & t \\ 0 & 0 & 1 \end{pmatrix}.$$

Define A, B, and C as follows:

$$A = \begin{pmatrix} 1 & 1 & 0 \\ 0 & 1 & 0 \\ 0 & 0 & 1 \end{pmatrix}, \quad B = \begin{pmatrix} 1 & 0 & 0 \\ 0 & 1 & 1 \\ 0 & 0 & 1 \end{pmatrix}, \quad C = \begin{pmatrix} 1 & 0 & 1 \\ 0 & 1 & 0 \\ 0 & 0 & 1 \end{pmatrix}.$$

One can check for all $k, l, m \in \mathbb{Z}$, the following holds:

$$A^k = \begin{pmatrix} 1 & k & 0 \\ 0 & 1 & 0 \\ 0 & 0 & 1 \end{pmatrix}, \quad B^l = \begin{pmatrix} 1 & 0 & 0 \\ 0 & 1 & l \\ 0 & 0 & 1 \end{pmatrix}, \quad C^m = \begin{pmatrix} 1 & 0 & m \\ 0 & 1 & 0 \\ 0 & 0 & 1 \end{pmatrix}.$$

Each element of H is then equal to one of the matrices:

$$A^k B^l C^m = \begin{pmatrix} 1 & k & m + kl \\ 0 & 1 & l \\ 0 & 0 & 1 \end{pmatrix}.$$

It is straightforward to check that

$$[A, B] = C, \quad [C, A] = [C, B] = 1.$$

To conclude this section, we would like to introduce the notion of *local* and *virtual* properties. These notions are very important when considering nilpotency, freeness, and finiteness.

DEFINITION 1.38. *Let P be a certain property of groups. Then a group G is said to be* locally P *if every finitely generated subgroup of G possesses the property P.*

DEFINITION 1.39. *In the previous notation, a group G is said to be* virtually P *if there is a subgroup $H \leq G$ of finite index for which the property P holds.*

CHAPTER 2

Algorithmic Problems in Group Theory

2.1. Introduction

The main purpose of this chapter is to introduce classical group-theoretic algorithmic problems. The problems introduced are relevant in the context of group-based cryptography. We begin with introducing combinatorial group theory, as this lays the foundation for the study of algorithmic problems in groups. We then present the notion of *Decision Problems*: problems where given a property $\mathcal{P}$ and an object $\mathcal{O}$, it is needed to decide if the object $\mathcal{O}$ has the property $\mathcal{P}$. Then, we present the search versions of some these problems: given a property $\mathcal{P}$ and a certain object $\mathcal{O}$ for which the property $\mathcal{P}$ is true, find a material witness to the property $\mathcal{P}$. For example, given two conjugate elements of a group, find a conjugator. In addtion, we discuss some algorithmic problems which have both graph-theoretic and group-theoretic analogues. Lastly, discuss the computability issues related to the possible solutions of these problems. For the interested reader, we refer [**51**] for the study of algorithmic problems in group theory. We refer to [**150**] for detailed information about combinatorial group theory, and to [**28**] for a historical perspective of the subject.

2.2. Combinatorial group theory

Combinatorial group theory is an essential topic in the present book as it is frequently used in group-based cryptography. The notion of a free group, which we will soon describe, will be ubiquitous in the text. We begin with the following definition required to define a free group.

DEFINITION 2.1. *Let X be a set called the* alphabet, *whose elements will be called* letters. *A* word *in X is defined as a finite sequence of letters and their formal inverses. The empty word, which will be denoted by ϵ, corresponds to the empty sequence.*

If a word w does not contain a subword of the form xx^{-1} or $x^{-1}x$ for any $x \in X$, w is called a *reduced* word. The set of all reduced words in the alphabet X is denoted $F(X)$.

EXAMPLE 2.2. *Let $X = \{x, y\}$ be an alphabet. Then x and y are letters of X and*

$$xy, \qquad y, \qquad xyxyx, \qquad x^{-1}yxy^{-1}, \qquad \textit{and} \qquad xy^{-1}xx^{-1}y$$

are examples of words over X. In particular, the first four words are examples of reduced words while the fifth is not.

Now that the necessary preliminary definitions have been stated, we are ready to define a free group.

DEFINITION 2.3. *The set $F(X)$ is a group under juxtaposition of letters, and is called the* free group *on $|X|$ generators.*

Observe that every element of $F(X)$ can be written uniquely as a product of elements of X and their inverses. Moreover, the free group is characterized by the cardinality of the set X; that is, if $|X| = n$, $F(X)$ is usually denoted by F_n and is called the free group of rank n.

The simplest example of a free group is the infinite cyclic group. It is a free group of rank 1, and it is clearly abelian. When $n \geq 2$ the free groups F_n are always nonabelian.

2.2.1. Presentations. A group presentation allows a group to be represented as a quotient of a free group. Before formally defining a presentation, we need the following definition.

DEFINITION 2.4. *Let G be a group and X a subset of G. Then the* normal closure, *denoted $\langle X\rangle^G$, of X in G is the smallest normal subgroup of G that contains X.*

DEFINITION 2.5. *Let $F(X)$ be a free group over an alphabet X, R a set of words and their inverses in the alphabet X, and $G = F(X)/\langle R\rangle^{F(X)}$. Then $\langle F \mid R\rangle$ is a* presentation *of G by generators and relations. The elements of $\langle R\rangle^{F(X)}$ are called* relations *or* relators.

It is easy to establish that every group G admits a presentation by generators and relations. To see this, let the set of generators X be all the elements of the group and let the relations be the elements of the kernel of the natural epìmorphism $F(X) \to G$ that identifies X.

EXAMPLE 2.6. *Here we provide a few examples of presentations of groups.*

- *The free group $F(X)$ generated by X of rank $|X|$ admits the presentation $\langle X | \;\rangle$.*
- *The cyclic group of order $n \in \mathbb{N}$ admits the following presentation:*

$$\mathbb{Z}_n = \langle x \mid x^n\rangle.$$

- *A free abelian group generated by a and b has the following presentation:*

$$\langle a, b \mid a^{-1}b^{-1}ab\rangle.$$

The last two examples are instances of the family of 1-relator groups.This family of groups as well as finitely presented groups will be relevant throughout the text. We define both of these families below.

DEFINITION 2.7. *Let G be a group presented by $\langle X \mid R\rangle$. If R has just one element, G is called a 1-relator group.*

DEFINITION 2.8. *If G admits a presentation $\langle X \mid R\rangle$ with $|X|$ and $|R|$ finite, then G is said to be finitely presented.*

It is straightforward to see that finite groups, cyclic groups and free groups of finite rank are examples of finitely presented groups. In general, every finitely presented group is finitely generated, but the converse is not always true.

REMARK 2.9. If a group G is defined by a presentation $\langle X \mid r_1, r_2, \ldots\rangle$ with $r_i \in R$ relators, it is customary to write $\langle X \mid r_1 = 1, r_2 = 1, \ldots\rangle$ since the relators are trivial in $G = F_X/R$. We will use both notations throughout the text.

2.3. Decision problems in groups

In this section, we introduce several important decision problems. The first decision problems were introduced by Max Dehn in 1911. These problems had an immense impact on algorithmic group theory, and still influence current research.

Dehn's decision problems: Let G be a group.

- *Word problem*: For any $g \in G$, determine if g is the identity element of G.
- *Conjugacy problem*: For any $x, y \in G$, determine if x and y are conjugate.
- *Isomorphism problem*: If G and G' are groups given by finite presentations, determine if G is isomorphic to G'.

2.3.1. Epimorphism problem. Let R_1 and R_2 be two finite group presentations. Determine if there exists an algorithm that decides if the group with presentation R_1 is a homomorphic image of the group defined by the presentation R_2.

2.3.2. Power decision problem. Let G be a finitely generated group. Given $x, y \in G$, determine whether there exists $n \in \mathbb{Z}$ such that $y = x^n$. Notice that this is equivalent to deciding whether $y \in \langle x \rangle$, and thus is the analogue of the discrete logarithm problem (see Section 2.6.1). A generalization of the power decision problem is the subgroup membership problem.

2.3.3. Subgroup membership decision problem. Let G be a group and let $H \leq G$. The subgroup membership decision problem, also known as the *generalized word problem*, asks if given any $g \in G$ whether or not $g \in H$. This problem can be generalized to the *rational subset membership problem*. Given a group G, the class of rational subsets of G is the smallest class that contains all finite subsets of G and that it is closed with respect to union, product and taking the free monoid generated by a set (i.e., using the Kleene star operation). The rational subset membership problem asks whether given a rational subset H of G and an element $g \in G$, we have $g \in H$.

2.3.4. The endomorphism decision problem. Let F be a free group and $x \in F$. The endomorphism decision problem asks whether there exists an algorithm that given $y \in F$ decides if there exists an endomorphism ϕ of F sending x to y.

2.3.5. The group homomorphism problem. Consider two finitely presented groups $G_1 = \langle S_1 | R_1 \rangle$ and $G_2 = \langle S_2 | R_2 \rangle$. The group homomorphism problem asks if there is a homomorphism $G_1 \to G_2$ which sends generators in S_1 to generators in S_2.

2.3.6. The n-th root problem. For $n \in \mathbb{N}$, an element x in a group G is a n-th root of an element a if the equation $x^n = a$ has solution. Given an element $g \in G$, the n-th root problem asks to determine if g has any n-th root. If the root of every element of G belongs to G, the group is said to be *complete*.

If the group G is under addition, the n-th root problem is equivalent to the following. Does there exist an algorithm such that for any $n \in \mathbb{N}$ and every element $h \in G$ that determines whether the equation $nx = h$ has at least one solution in G.

2.3.7. Subgroup intersection problem. Let G be a finitely generated group, and $H, L \leq G$ be two finitely generated subgroup of G. The *subgroup intersection problem* is to determine whether the subgroup $H \cap L$ is also finitely generated. If this holds, a group is said to satisfy Howson's property.

2.4. Search problems

In this section we introduce some group theoretic search problems as these problems are essential in the study of noncommutative cryptography. Several protocols, including the Ko-Lee key exchange and the Anshel-Anshel-Goldfeld key exchange (see Chapter 4), are based a group theoretic search problem. We state the relevant search problems below.

Once again, we let G be a group given by the presentation $\langle X|R\rangle$ and every element of G is given as a product of generators in $X^{\pm 1}$.

2.4.1. Word search problem. Given a finitely presented group G and an element $g = 1$ in G find a presentation of g as a product of conjugates of defining relators and their inverses.

2.4.2. Conjugacy search problem. Let G be a group. Consider group elements $a_1, \ldots a_n, b_1, \ldots b_n \in G$ with a_i conjugate to b_i for $i = 1, \ldots, n$. The simultaneous conjugacy search problem is to find $c \in G$ such that $a_i^c = b_i$ for $1 \leq i \leq n$. The case where $n = 1$ is the conjugacy search problem.

2.4.3. Power conjugacy search problem. Let G be a group. Find for some $x, y \in G$, an element $g \in G$, and $n \in \mathbb{N}$ such that $x^n = y^g$. Note that for $n = 1$ this reduces to the conjugacy search problem, whereas if g is the identity in the group this reduces to the power search problem.

2.5. Algorithmic problems in graphs and groups

In this section we present algorithmic problems which have both graph theoretic and group theoretic versions.

2.5.1. Geodesic length problem. Let $G = \langle X \rangle$ be a group generated by a set X. Denote $|w|$ as the length of a word in the alphabet $X^{\pm 1}$, and ρ the canonical epimorphism of $F(X)$ onto G, where $F(X)$ denotes the free group over the set X. The geodesic length $l_X(g)$ of an element $g \in G$ with respect to X is

$$l_X(g) = \min\{|w| \mid w \in F(X),\ \rho(w) = g\}.$$

Depending on the context we will also use the notation $|g|_X$ instead of $l_X(g)$.

We say that a word $w \in F(X)$ is geodesic if $|w| = l_X(\rho(w))$. The geodesic length problem states the following: given a word $w \in F(X)$, find $l_X(\rho(w))$.

2.5.2. Subgroup distortion. Let G be a group generated by a finite set S, and let H be a subgroup of G generated by a finite set T. The *distortion* of the subgroup H in G is a map

$$\begin{aligned} Dist_G(H) &: \mathbb{N} \to \mathbb{N} \\ &n \mapsto \max\{l_T(h) \mid l_S(h) \leq n\}. \end{aligned}$$

It is clear that this map depends on the generating sets S and T. Although the map is dependent on the generating sets, it should be noted that two finite generating sets have equivalent distortion functions; that is, the functions differ by multiplicative constants. The degree of failure of this inequality is used to define the distortion.

2.5.3. The decomposition problem. Recall that a graph J is a *join* of two subgraphs $J \cong J_1 * J_2$ if $V(J) = V(J_1) \cup V(J_2)$ and if for each pair $v \in V(J_1)$ and $w \in V(J_2)$, we have $\{v, w\} \in E(J)$. A join $J_1 * J_2$ is called nontrivial if both J_1 and J_2 are nonempty. A join decomposition $J = J_1 * \cdots * J_n$ of a graph J is maximal if for each i, the graph J_i does not decompose as a nontrivial join. A maximal join decomposition of a finite simplicial graph is unique.

The *decomposition problem* states: given a graph Γ decompose Γ as a join of simpler graphs. Here the notion of simpler means that the join factors have fewer vertices than Γ.

2.5.4. The clique problem. A *k-clique* is a complete graph on k vertices. The clique problem states, given a graph Γ, find the induced subgraphs of Γ that are k-cliques for a certain k. There are many different variations of this problem. For instance, find a maximal (with respect to inclusion) k-clique inside Γ; list all k-cliques; check if, given k, there exists a j-clique in Γ for some $j > k$.

2.5.5. The independent set problem. An *independent set* inside a graph Γ is a set of vertices of Γ such that there is no edge in the subgraph of Γ which they span. The independent set problem asks whether or not a maximal (with respect to inclusion) independent set inside the graph Γ can be found.

2.5.6. Induced graphs. The previous questions belong to a large family of problems concerning induced graphs by subsets of the set of vertices of a graph. One such interesting problem is the *induced subgraph isomorphism problem*, which given a pair of graphs Γ and Γ', consists of finding a subgraph of Γ which is isomorphic to Γ'. Observe that in the special case Γ' is a clique, we obtain the clique problem, whence we conclude that this problem is in general NP-complete. In the case where Γ' has no edges, we obtain the independent set problem. Another interesting special case is the *snake-in-the-box problem*, in which Γ' is a chain (i.e., a connected graph with two vertices of degree one and the remaining vertices of degree two) and in which Γ is an hypercube (see [**128**]).

2.5.7. Subdivision problem. Recall that given a graph Γ, a subdivision of Γ is a graph obtained by performing successive subdivisions on the edges. The problem in this case is, given graphs Γ and Γ', find Γ' as a subgraph of a subdivision of Γ.

2.5.8. Graph coloring. A *(vertex) coloring* of a graph is an assignment of colors to the vertices of a graph in such a way that no two adjacent vertices are assigned the same color. The *chromatic number* of a graph is the minimum number of colors that are necessary to color the graph. Another variation of this is the dual *edge coloring*, which is an assignment of colors to the edges of a graph in such a way that no two edges incident to a common vertex are assigned the same color.

2.5.9. Vertex cover problem. Given a graph Γ, a *vertex cover* is a subset V' of the set of vertices V such that every edge of Γ is incident with at least one vertex of V'. Given a certain $k > 0$, the *vertex cover problem* is the problem of deciding if there exists a vertex cover of Γ with exactly k vertices. This problem is known to be NP-complete, even for planar graphs, although it is fixed-parameter tractable. A natural variation is the minimum vertex cover problem, which consists of finding the minimum k for which there exists a vertex cover with k vertices. This problem is known to be NP-hard.

2.5.10. Arboricity. Recall that a *forest* is an acyclic graph, i.e., a graph whose connected components are trees. The *arboricity* of a graph Γ is defined as the minimum number k such that there exists k subgraphs of Γ that are forests and whose union contains all the edges of Γ. The problem of arboricity is to find such a k. This can be solved in polynomial time.

2.6. Other problems in groups

2.6.1. The discrete logarithm problem. Let G be a finite group. Given $x, y \in G$, the discrete logarithm problem (denoted DLP) is to find a positive integer a such that $x^a = y$ (if it exists). This is usually defined in the setting of cyclic groups because the discrete logarithm exists for all elements and all nontrivial bases.

The DLP can be generalized to several components as follows. Let $\mathbf{x} = (x_1, \ldots, x_n)$ be a tuple of elements such that $G = \langle x_1, \ldots, x_n \rangle$. Given $y \in G$, the "generalized" DLP of y with respect to $\mathbf{x}$ is to find a_i such that y can be written uniquely as $\mathbf{x}^{\mathbf{a}} = (x_1^{a_1}, \ldots, x_n^{a_n}) = y$ where $0 < a_i < |x_i|$ for every i.

2.6.2. Degree of n-nilpotency. Roughly speaking, the degree of n-nilpotency measures how close a group is to being nilpotent of nilpotency class n. In particular, this is a way to measure how close Engel groups are to being nilpotent. This is relevant for us since some protocols work better using Engel (or nilpotent) groups (see Chapter 13).

In [**156**] the degree of n-nilpotency ($n \geq 2$) of a finite group G has been defined as follows:

$$d^{(n)}(G) = \frac{|\{(x_1, \cdots, x_{n+1}) \in G^{n+1} \mid [x_1, \cdots, x_{n+1}] = 1\}|}{|G|^{n+1}}.$$

For ease of notation write $d(G) = d^{(1)}(G)$. Clearly $d(G) = 1$ if and only if the group G is abelian. In [**95**] Gustafson proved that if $d^{(1)}(G) > 5/8$, then G is abelian. Moreover if $d(G) > 1/2$, then G is nilpotent [**142**].

2.6.3. Learning with errors. The learning with errors (LWE) problem was introduced by Regev in [**182**], and has become one of the most known problems in lattice-based cryptography. It has been used to construct several cryptosystems, and it is believed to be hard even for quantum computers.

Informally speaking, the problem of learning with errors is to *learn* a particular function by sampling the input/output behavior if some of the outputs are incorrect. For more information one can see [**12**].

In the following, we define the learning homomorphism problem with noise (denoted LHN) which is a generalization of the LWE. We will denote the set of homomorphisms from a group G to a group H with $Hom(G, H)$.

Suppose that one is able to efficiently sample from a probability distribution α over the group G. Let $\varphi \in Hom(G, H)$, where G and H are finitely generated groups, and let $g_1, \dots g_m \in G$.

Denote with $P_\eta(\phi, \alpha)$ the probability distribution over $G \times H$ obtained by sampling an element $g \in G$ according to $h, \eta \in H$ according to α, which gives $(g, \varphi(g)h)$ as output.

Let α and β be probability distributions over G and H, respectively. We say that an algorithm solves the LHN with noise for G, H, α and β if for any $\varphi : G \to H$, the algorithm is able to learn φ given a set of samples from the distribution $P_\eta(\phi, \alpha)$ with high probability. For a detailed reference see [**12**] and [**56**].

CHAPTER 3

Classical Cryptography

3.1. Introduction

In this chapter, we introduce several protocols in classical cryptography, some of which have noncommutative analogs. The concepts on which these protocols are based on are essential in understanding cryptography — both symmetric and asymmetric. We begin with the foundations of symmetric key cryptography. We follow the exposition of [**107**]. The typical scenario under consideration consists of two parties, Alice and Bob, who would like to communicate with one another privately and Eve, the adversary, who is attempting to gain information about the private communications. Suppose Bob would like to send a secret message, m, to Alice. Bob uses a secret key, k, to encrypt the plaintext message, m, to a ciphertext, c. After receiving the ciphertext, c, Alice uses the same key, k, to decrypt the ciphertext c to reconstruct m. For Bob to encrypt his message and Alice to decrypt the message they must both know the secret key k, which is the key feature in symmetric key cryptography. In order to provide security it must be the case that the adversary, Eve, does not know k, cannot easily guess k and cannot recover the message from the ciphertext without k.

The encryption and decryption processes can be formalized by viewing the processes as functions. Let $\mathcal{C}$ and $\mathcal{M}$ denote the space of possible ciphertexts and the space of possible plaintexts, respectively. Then one can consider encryption as a function e_k dependent on a key k

$$e_k : \mathcal{M} \to \mathcal{C}.$$

Similarly, one can consider decryption as a function d_k dependent on a key k

$$d_k : \mathcal{C} \to \mathcal{M}.$$

These functions have the property that $d_k(e_k(m)) = m$; that is, given a ciphertext the function d_k will recover the plaintext.

In this scenario, it is generally assumed that the adversary will have knowledge of the encryption method being implemented, but not of the key k. Hence, the security of a cryptosystem should depend only on the secrecy of the key rather than the secrecy of the encryption algorithm. This principle is referred to as *Kerckhoff's principle.*

Next we consider the scenario where Alice and Bob would like to exchange information, but do not necessarily have the opportunity to communicate over a secure channel to determine a key k. In 1976, Whitfield Diffie and Martin Hellman published their seminal paper (see [**43**]) in cryptography "New Directions in Cryptography", which addresses this situation. This paper introduced the concept of asymmetric cryptography, which is commonly referred to as public-key cryptography, a concept based on one-way functions and trapdoor information. The basic idea of a one-way function is that it is easy to compute and hard to invert, unless you have special information. We can define encryption and decryption functions as in the case of symmetric key cryptography with one change; rather than having only one key, there is a public key and a private key. We denote the public key k_1 and the private key k_2. In order to implement an encryption algorithm the public key is used, giving us the encryption function

$$e_{k_1} : \mathcal{M} \to \mathcal{C}.$$

The decryption function

$$d_{k_2} : \mathcal{C} \to \mathcal{M}$$

is dependent on the private key. This allows anyone to perform encryption and only those with the private key to perform decryption.

3.1.1. RSA public-key cryptosystem. The RSA public-key cryptosystem, invented by Ron Rivest, Adi Shamir, and Leonard Adleman (RSA) in 1977, relies on the difficulty of factoring the product of two large prime numbers. RSA is widely used for sharing sensitive data over insecure channels. Suppose Alice would like to encrypt a message m to send to Bob. The RSA scheme used to encrypt m can be summarized as follows (see [**185**]).

(a) Two large primes, p and q, are randomly chosen by Bob. These values are kept secret.
(b) Bob computes $N = pq$ and $(p-1)(q-1)$. He then chooses encryption exponent e with $\gcd(e,(p-1)(q-1)) = 1$. The pair (N, e) is made public.
(c) Using Bob's public key, (N, e), Alice computes $c \equiv m^e \mod N$ and sends c to Bob.
(d) To recover the message, Bob begins by computing d satisfying $ed \equiv 1 \mod (p-1)(q-1)$. Then, he computes $c^d \mod N$. Note that $c^d \mod N \equiv (m^e)^d \mod N \equiv m^{ed} \mod N \equiv m \mod N$.

The security of this scheme relies on the fact that it is easy for one to determine d satisfying $ed \equiv 1 \mod (p-1)(q-1)$ knowing p and q, but it is difficult without this additional information. By having knowledge of d, one can solve the congruence $m^e \equiv c \mod N$ for m as noted above; that is, one can find the eth root mod N.

Here, we provide a small example of the implementation of an RSA cryptosystem.

EXAMPLE 3.1. *Suppose Alice would like to send the message* 718389 *to Bob over an open channel. The RSA encryption scheme can be implemented as follows.*

(a) *Bob randomly chooses primes* $p = 9461$ *and* $q = 11383$.
(b) *Bob then computes* $N = 107694563$ *and* $(p-1)(q-1) = 107673720$. *Encryption exponent* $e = 573877$ *with* $gcd(573877, 107673720) = 1$ *is chosen. Bob's public key is* $(107694563, 573877)$.
(c) *Using Bob's public key, Alice computes* $46772882 \equiv 718389^{573877} \mod 107694563$.
(d) *Bob solves* $573877d = 1 \mod 107673720$ *to get* $d = 61005733$. *He then computes* $46772882^{61005733} \equiv 718389 \mod 107694563$.

3.1.2. Diffie-Hellman key exchange. The Diffie-Hellman key exchange is based on the discrete logarithm, and it exploits the difficulty of computing discrete logarithms in certain groups. For convenience, we restate the discrete log problem here.

DEFINITION 3.2. *Let G be a group with elements g and h. The* discrete log problem *for G consist in finding an integer n such that*

$$g^n = h.$$

Originally, the Diffie-Hellman key exchange was based on the discrete log problem in $\mathbb{Z}_p^*$, the group of nonzero integers modulo a prime p under multiplication. We present the classical Diffie-Hellman scheme below; see [**69**]:

(a) Alice and Bob agree on a large prime p, and a primitive root g modulo p. These need not be kept secret and can be shared by a group of users.
(b) Alice chooses a large random integer x, which is kept secret, privately computes $X = g^x \bmod p$, and sends X to Bob by the possible insecure channel.
(c) Similarly, Bob chooses a large random integer y, which is kept secret, privately computes $Y = g^y \bmod p$, and sends Y to Alice across the possible insecure channel.
(d) Then Alice privately computes $k = Y^x \bmod p$, and Bob symmetrically computes $k' = X^y \bmod p$.

Note that $k = k' \bmod p$, as

$$k' = X^y = (g^x)^y = g^{xy} = (g^y)^x = k \bmod p,$$

and thus $k = k'$ is the common secret of Alice and Bob. Since Alice and Bob have the additional information x and y, respectively, the secret is easily computable for them. Without this information, no one else on the network can readily determine k.

Clearly, if one can solve the discrete log problem in $\mathbb{Z}_p^*$ then he or she can obtain the shared key. This is one method of an adversary obtaining the shared key. It should be noted that an adversary can also obtain the shared key by solving another problem, the Diffie-Hellman problem.

DEFINITION 3.3. *Let p be a prime number and g a primitive root of $\mathbb{Z}_p^*$. The* Diffie-Hellman problem *is to find the value $g^{xy} \mod p$ given the values $g^x \mod p$ and $g^y \mod p$.*

It is clear if one can solve the discrete log problem then one can solve the Diffie-Hellman problem. It is not known if an algorithm to solve the Diffie-Hellman problem will provide an algorithm to solve the discrete log problem.

Now we present a small example of the Diffie-Hellman key exchange protocol.

EXAMPLE 3.4. *Suppose that Alice and Bob agree upon the large prime $p = 9719$ with primitive root $g = 17$. The scheme is implemented as follows.*

(a) *Alice chooses random integer 1247 and computes $17^{1247} \mod 9179 = 7259$. 7259 is public information.*
(b) *Bob chooses random integer 4562 and computes $17^{4562} \mod 9179 = 7780$. 7780 is public information.*

(c) *Alice computes* $7780^{1247} \mod 9179 = 4782$.
(d) *Bob computes* $7259^{4562} \mod 9179 = 4782$.

In this scenario the shared key is 4782, which is computable for both Alice and Bob.

3.1.3. ElGamal encryption. By making a slight modification to the Diffie-Hellman key exchange, Taher ElGamal created his encryption scheme in 1985 (see [**50**]). In this scenario, we suppose that Alice would like to send a message, m, to Bob over an open channel. Bob begins by making the following information public: a large prime p, a primitive root $g \mod p$, and $Y = g^y \mod p$, where y, a large random integer, is Bob's secret key.

Using Bob's public key, Alice encrypts the message m as follows.

(a) Alice chooses a large random integer x, which she keeps secret, and computes $X = g^x \mod p$.
(b) Alice computes the shared key, $Y^x \mod p$.
(c) Alice then encrypts the message m by computing $M = mY^x \mod p$.
(d) Finally, Alice sends Bob the encrypted message M and the header X.

Bob has the ability to decrypt Alice's message using his secret key y by following the following steps.

(a) Bob uses his secret y to compute $X^y = (g^x)^y = (g^y)^x = Y^x \mod p$.
(b) Knowing $Y^x \mod p$ and p, Bob can then compute $(Y^x)^{-1} \mod p$.
(c) Bob can now decrypt the message by computing $M(Y^x)^{-1} = mY^x(Y^x)^{-1} = m \mod p$.

Hence, in order to decrypt the message it was required to know Bob's secret, y, but not Alice's secret, x. As with the Diffie-Hellman key exchange, the scheme relies on the difficulty of solving the discrete log problem. Unlike the Diffie-Hellman key exchange, the ElGamal scheme is intended to encrypt messages; although, asymmetrical encryption is not computationally efficient.

Here, we provide an example of the implementation of the ElGamal encryption.

EXAMPLE 3.5. *Suppose that Bob publicizes the following information: prime* $p = 10391$, *primitive root* $g = 19$, $Y = 3886$. *Note that* $3886 = 19^{2391} \mod 10391$ *and* 2391 *is Bob's secret key.*
Alice would like to send the message, $m = 9999$, *to Bob.*
Alice randomly chooses her secret 1397, and uses this to compute the shared key

$$Y^x = 3886^{1397} \mod 10391 = 6865$$

and

$$X = 19^{1397} \mod 10391 = 1147.$$

She then encrypts her message $M = 9999(6865) \mod 10391 = 189$. *She sends the ciphertext* 189 *to Bob along with the header* $X = 1147$.
In order to decrypt the message, Bob uses the header to determine the shared key by computing $X^y = 1147^{2391} \mod 10391 = 6865$. *He then computes the inverse of* $6865 \mod 10391$, *which is* 5001. *Finally, Bob recovers the message by computing*

$$m = 189(5001) \mod 10391 = 9999.$$

3.2. Digital signature schemes

Consider the scenario of applying to graduate school. Typically one must provide letters of recommendation in the application materials. When mailing a physical copy of the letter, the convention has been to sign the seal on the envelope. This action ensures that the letter has been unaltered and that the letter was in fact written by the individual listed as reference. A digital signature scheme serves the same purpose as signing the seal on a handwritten letter; it should verify that the message received is from the correct sender and that the message has not been altered.

More formally, a digital signature scheme is defined as a triple of algorithms (G, σ, V) such that

- G is a key generation algorithm which produces the pairs (P, S), where P is called a public key and S is called a secret key. The secret key S is a private signing key; while the public key P is a public verification key.
- σ is a signing algorithm which when given a secret S and a message m produces an output s which we call the signature of m.
- V is a verification algorithm which when given a public key P, a digital signature s, and a message m, determines whether or not the signature is valid. The verification algorithm accepts all valid signatures and rejects all invalid signatures.

To ensure security of a digital signature scheme, it is necessary that one can create valid signatures using the private key S, but knowledge of the public key P does not reveal the private key. Without this requirement, anyone can sign the document. To ensure this property a digital signature scheme must possess the following properties:

- Given a public key P an adversary cannot feasibly determine the secret key S. Additionally, the adversary cannot determine any other private key that will produce the same signature as S.
- Given a public key P and a list of signed documents with their signatures, an adversary cannot feasibly determine a valid signature on any document not in the list.

Every time a digital signature scheme is implemented it signs a new document. Hence, every time a new signature and document pair is revealed, new information is given to the adversary. The second condition is stating that the adversary gains no new information other than the signature and document pair (see [**107**]).

3.2.1. RSA digital signature scheme.

The RSA digital signature scheme was proposed in [**185**] in the same article where Rivest, Shamir, and Adleman first proposed the RSA encryption scheme. Although the scenario considered here is different, the initial steps of the scheme are the same as in the encryption scheme. In this scenario Bob would like to digitally sign a document D in such a way that any recipient can verify Bob's signature. The scheme can be implemented as follows.

Public-key generation:

- Two large primes p and q are randomly chosen by Bob. These are kept secret.
- Bob computes and publicizes $N = pq$ and verification exponent v with $\gcd(v, (p-1)(q-1)) = 1$.

Signing algorithm:

- Bob computes the signing exponent s by solving
$$sv \equiv 1 \mod (p-1)(q-1).$$
- Bob can then sign a document D, with $1 < D < N$, by computing $S = D^s \mod N$.

Verification algorithm:

- In order to verify a signature, Alice can compute $S^v \mod N$ and check that it is equal to D.

As with the RSA encryption scheme, if an adversary can factor N then the adversary can determine the private signature key s, allowing them to forge signatures on any given document. By having the additional knowledge of p and q, the adversary can solve the congruence $x^v \equiv D \mod N$ and send the signed document x.

Here, we present a small example of the RSA digital signature scheme.

EXAMPLE 3.6. *Suppose that Bob would like to sign a digital document* $D = 10143$. *In order to sign the document, Bob begins by randomly choosing two primes* $p = 7151$ *and* $q = 9137$. *Then,*

(a) *Bob computes and publicizes* $N = 7151 * 9137 = 65338687$ *and verification exponent* $v = 37078139$. *Note that* $\gcd(37078139, 65322400) = 1$, *where* $7150 * 9136 = 65322400$.

(b) *Bob now computes signing exponent* $s = 10015059$ *by solving* $37078139s \equiv 1 \mod 65322400$.

Bob can now sign the document by computing $S = 17188707 \equiv 10143^{10015059} \mod 65338687$.

In order to verify the signature, Alice uses Bob's public verification exponent $v = 37078139$ *and computes*

$$S^v = 17188707^{37078139} \mod 65338687 \equiv 10143 \mod 65338687.$$

Since $S^v \equiv D \mod N$, *Alice verified Bob's signature on the document* D.

3.2.2. ElGamal signature scheme. The ElGamal digital signature scheme was proposed in [**50**] in 1985. Like the ElGamal encryption scheme, this digital signature scheme is based on the discrete log problem. Suppose that Bob would like to send and sign a digital document D in such a way that any recipient can verify Bob's signature. The scheme can be implemented as follows.

Public-key generation:

- Bob begins by randomly choosing a large prime p, primitive root $g \mod p$ and random integer x. Here x is Bob's secret key. Both the prime p and primitive root g are made public.
- Bob computes and publicizes $X = g^x \mod p$.

Signing algorithm

- Bob can now sign document D by determining a pair of integers (r, s) with $0 \neq r, s \neq p-1$ and $g^D = X^r r^s \mod p$.

Verification algorithm:

- In order to verify the signature one must check that $g^D = X^r r^s \mod p$ holds.

In order to generate a pair (r, s) which constitutes a signature, the signer begins by choosing a random number k such that $0 \neq k \neq p-1$ and $\gcd(k, p-1) = 1$.

Let $r = g^k \bmod p$. Now we want to compute an s such that $g^D = y^r r^s \equiv g^{xr+ks}$ mod p. In terms of the exponents, this relationship is $D \equiv xr + ks \pmod{p-1}$. Hence $s = (D - xr)k^{-1} \bmod p - 1$. The signature of D is the pair (r, s). Clearly, if an attacker could solve the discrete logarithm problem, he could break the scheme completely by computing the secret key x from the information in the public file. Moreover, the random number k should be difficult for an attacker to determine as this will allow him to solve the discrete log problem. To see this, note that given a digital document D and signature (r, s) the congruence $D \equiv xr + ks \mod p - 1$ is solvable for the secret key x.

Here, we present an example of the ElGamal digital signature scheme.

EXAMPLE 3.7. *Suppose Bob would like to sign and send the digital document* $D = 1079$.

Public-key generation:

- *Bob begins by choosing prime* $p = 71327$ *and primitive root* $g = 571$.
- *Bob chooses a secret integer* $x = 26723$ *and computes* $33697 \equiv 571^{26723} \mod 71327$. *Bob publicizes* $X = 33697$, $p = 71327$, $g = 571$.

Signing algorithm:

- *Bob randomly chooses* $k = 29095$ *with* $gcd(29095, 71326) = 1$.
- *He computes* $r = 571^{29095} \equiv 56 \mod 71327$ *and* $k^{-1} \equiv 10637 \mod 71326$.
- *Lastly, Bob determines* $s = 31031 \equiv (1079 - 26723 * 56)10637 \mod 71326$.
- *The signature for* $D = 1079$ *is the pair* $(56, 31031)$.

Verification algorithm:

- *Upon receiving* $D = 1079$ *and* $(56, 31031)$, *one may verify the signature by computing* $g^D \mod p$ *and* $X^r r^s \mod p$ *and checking they are equal. Since*

$$g^D = 571^{1079} \equiv 48333 \mod 71327$$

and

$$X^r r^s = 33697^{56} 56^{31031} \equiv 48333 \mod 71327$$

are equal, the signature is valid.

3.3. Secret sharing schemes

To introduce the idea of secret sharing schemes, consider the scenario of keeping the money in a bank vault secure. For simplicity, suppose there are three bank managers. In order to secure the money, keys can be distributed to each of the three managers in such a way that two of the three keys must be utilized to unlock the vault. This would allow for additional security since at least two of the managers must be corrupt in order to open the vault, and no single manager gains access to the safe. This idea can be generalized to any number of managers requiring a minimum number of them to gather in order to open the vault. Secret sharing schemes are the analog of this scenario in the digital word. More formally, secret sharing schemes provide a method to distribute a secret, K, among n persons in such a way that any t, where $0 < t \leq n$, of the participants can recover the secret, but no $t - 1$ participants can. A method that allows one to do this is called a (t, n)-threshold scheme.

3.3.1. Shamir's secret sharing scheme. Shamir created a (t,n)-threshold scheme in 1979 that uses polynomial interpolation. In Shamir's scheme the secret is an element $x \in \mathbb{Z}_p^*$, where p is a large prime. In order to distribute the secret, the dealer begins by choosing a polynomial f of degree $t-1$ such that $f(0)=x$. Then he sends the value $x_i = f(i)$ secretly to participant P_i for $1 \le i \le n$. In order for the participants to recover the secret, they use polynomial interpolation to recover f and hence the secret $f(0)$. In this setting, no $t-1$ participants can gain any information about the secret while any t of them can (see [**200**]). To illustrate the idea, we provide a small example.

EXAMPLE 3.8. *Let $p = 21379$. Suppose the dealer would like to distribute the secret $K = 11291$ among 10 participants with a threshold of 7. The dealer begins by forming a polynomial of degree 6,*

$$f(x) = 4798x^6 + 19809x^5 + 3808x^4 + 7080x^3 + 12458x^2 + 15890x + 11291$$

with $f(0) = 11291$.
The dealer then distributes the following numbers to each participant as follows.

- $f(1) = 10997$ *to* P_1,
- $f(2) = 18344$ *to* P_2,
- $f(3) = 2840$ *to* P_3,
- $f(4) = 14258$ *to* P_4,
- $f(5) = 14899$ *to* P_5,
- $f(6) = 2323$ *to* P_6,
- $f(7) = 16347$ *to* P_7,
- $f(8) = 8931$ *to* P_8,
- $f(9) = 4532$ *to* P_9,
- $f(10) = 16451$ *to* P_{10}.

Now any subset of seven participants can pool their shares and recover the secret. For example, we will consider when P_1, P_2, P_5, P_6, P_7, P_9, and P_{10} pool their shares together. By pooling their shares, the participants form the following system of linear equations mod 11291:

$$\begin{aligned}
a_6(1)^6 + a_5(1)^5 + a_4(1)^4 + a_3(1)^3 + a_2(1)^2 + a_1(1) + a_0 &= 10997,\\
a_6(2)^6 + a_5(2)^5 + a_4(2)^4 + a_3(2)^3 + a_2(2)^2 + a_1(2) + a_0 &= 18344,\\
a_6(5)^6 + a_5(5)^5 + a_4(5)^4 + a_3(5)^3 + a_2(5)^2 + a_1(5) + a_0 &= 14899,\\
a_6(6)^6 + a_5(6)^5 + a_4(6)^4 + a_3(6)^3 + a_2(6)^2 + a_1(6) + a_0 &= 2323,\\
a_6(7)^6 + a_5(7)^5 + a_4(7)^4 + a_3(7)^3 + a_2(7)^2 + a_1(7) + a_0 &= 16347,\\
a_6(9)^6 + a_5(9)^5 + a_4(9)^4 + a_3(9)^3 + a_2(9)^2 + a_1(9) + a_0 &= 4532,\\
a_6(10)^6 + a_5(10)^5 + a_4(10)^4 + a_3(10)^3 + a_2(10)^2 + a_1(10) + a_0 &= 16451.
\end{aligned}$$

By solving the system, the participants obtain the coefficients of the polynomial; hence, also determining the secret. Rather than solving the system of equations, the participants can also use Lagrange's interpolation formula to recover f.

3.4. Multilinear maps

Boneh and Silverberg first introduced the notion of utilizing multilinear maps in cryptographic schemes in [**17**]. We begin with some definitions.

DEFINITION 3.9. *Let G_1 and G_2 be groups. A map $e : G_1^n \to G_2$ is an* n-multilinear map *if it satisfies the following properties:*
(a) *G_1 and G_2 both have prime order p;*
(b) *For $a_1 \ldots, a_n \in \mathbb{Z}$ and $g_1, \ldots, g_n \in G_1$ we have*

$$e(g_1^{a_1}, \ldots, g_n^{a_n}) = e(g_1, \ldots, g_n)^{a_1 \cdots a_n};$$

(c) *If $g \in G_1$ is a generator for G_1, then $e(g, \ldots, g)$ is a generator of G_2.*

REMARK 3.10. Property (c) is the definition of a map that is nondegenerate.

DEFINITION 3.11. *An n-multilinear map $e : G_1^n \to G_2$ is a* cryptographic n-multilinear map *if the following properties hold:*
(a) *The group operation in G_1 and G_2 are efficiently computable.*
(b) *The map e is efficiently computable.*
(c) *The discrete log problem in G_1 is hard.*

The construction of a cryptographic n-multilinear map for $n > 2$ remains an open problem. We would like to note that the third property ensures that the discrete log problem is also hard in G_2 since an efficiently computable n-multilinear map, $e : G_1^n \to G_2$, can be used to reduce the discrete log problem in G_2. To see this, consider elements $g, h \in G_1$ such that $h = g^x$. In order to solve the discrete log problem in G_1, one must determine the value x given the values h and g. One may reduce this problem in G_1 to a discrete log problem in G_2 by computing

$$y = e(g, g, \ldots, g)$$

and

$$z = e(h, g, \ldots, g).$$

By multilinearity, we have $z = e(h, g, \ldots, g) = e(g^x, g, \ldots, g) = e(g, g, \ldots, g)^x$. Hence, we have that $z = y^x$ in G_2. This means that if the discrete log problem is hard in G_1 then it must also be hard in G_2 (see [**17**]).

There are several applications of multilinear maps in cryptography including a one round $n + 1$-way Diffie-Hellman key exchange, building unique signatures and verifiable pseudo random functions, broadcast encryption, and indistinguishability obfuscation. Here we will present the notion of of a one round $n + 1$-way Diffie-Hellman key exchange. For the interested reader, we refer to [**17**].

In a one-round n-way Diffie-Hellman key exchange, $n + 1$ parties would like to determine a shared key and each party is only allowed to broadcast one value to the other parties. Once all the parties have broadcasted their values each participant should be able to compute the shared key. The scheme developed in [**17**] is as follows.

Let G_1 be a group of prime order p with generator g. Let $e : G_1^n \to G_2$ be a n-multilinear map. The map e, prime p and generator g are public information.
(a) Each participant P_i for $i = 1, \ldots n + 1$ chooses a random integer a_i with $1 < a_i < p - 1$ and computes $h_i = g^{a_i}$ in G_1. The integer a_i is kept secret, while h_i is published.
(b) Once each participant has published h_i, participant j can compute the shared key by determining

$$K = e(h_1, \ldots, h_{j-1}, h_{j+1}, \ldots, h_{n+1})^{a_j}.$$

Note that each participant will arrive at the same value since for each j the participant computes

$$\begin{aligned} K = e(h_1, \dots, h_{j-1}, h_{j+1}, \dots, h_{n+1})^{a_j} &= e(g^{a_1}, \dots, g^{a_{j-1}}, g^{a_{j+1}}, \dots g^{a_{n+1}})^{a_j} \\ &= (e(g, g, \dots, g)^{a_1 \dots a_{j-1} a_{j+1} \dots a_{n+1}})^{a_j} \\ &= e(g, g, \dots, g)^{a_1 \dots a_{n+1}}. \end{aligned}$$

The security of this scheme relies on the difficulty of the multilinear Diffie-Hellman assumption, which can be stated as follows.

DEFINITION 3.12. *The* multilinear Diffie-Hellman assumption *says that given* $g, g^{a_1}, \dots, g^{a_{n+1}}$ *in* G_1, *it is hard to compute* $e(g, g, \dots, g)^{a_1 \dots a_{n+1}}$ *in* G_2.

Hence, if the multilinear Diffie-Hellman assumption holds the $n+1$-way Diffie-Hellman key exchange above is secure for every $n > 1$.

3.4.1. Multilinear maps in nilpotent groups. Let G be a nilpotent group of class $n > 1$ and $g_1, \dots, g_n \in G$. One can easily prove by induction on n that for any $a_1, \dots, a_n \in \mathbb{Z}$ the following identity holds:

$$[g_1^{a_1}, \dots, g_n^{a_n}] = [g_1, \dots, g_n]^{\prod_{i=1}^n a_i}. \tag{3.1}$$

Hence if G is nilpotent, the map e

$$\begin{aligned} e : G^n &\to G \\ (g_1, \dots, g_n) &\mapsto [g_1, \dots, g_n] \end{aligned}$$

is a multilinear map. In addition, if we fix $x \in G$, we can construct another multilinear map f given by

$$\begin{aligned} f : G^{(n-1)} &\to G \\ (g_1, \dots, g_{n-1}) &\mapsto [x, g_1, \dots, g_{n-1}]. \end{aligned}$$

Notice that we want the multilinear map to be nondegenerate. This is the case if we assume that G is not $(n-1)$-Engel (see Chapter 13). Indeed, in this condition, there exists $g \in G$ such that $[x,_{(n-1)} g] \neq 1$, which implies that f is nondegenerate.

Part 2

Post-quantum Cryptography and Cryptanalysis

CHAPTER 4

Noncommutative Cryptographic Protocols

4.1. Noncommutative key exchange protocols

4.1.1. Introduction. With the advent of the quantum computer, a new interdisciplinary field arose: post-quantum cryptography. Computer scientists and mathematicians alike have been working diligently to ensure that the transfer of data in a post-quantum world will remain secure. Many key exchange protocols based on mathematical problems have been proposed in recent years. Here we present several key exchange protocols based on an underlying noncommutative structure. The security of each key exchange protocol is based on a well-known algorithmic problem.

We begin with three groundbreaking ideas in the field: AAG key exchange, Ko, Lee, et al. key exchange, and the noncommutative ElGamal key exchange. We then introduce a protocol based on a generalization of the Diffie-Hellman key exchange and two key exchange protocols based on a more complex underlying structure, the semidirect product. For the interested reader, we refer to the survey papers [**58**] and [**120**], as well as the book [**162**].

4.1.2. AAG key exchange. The AAG key exchange was first proposed by I. Anshel, M. Anshel, and D. Goldfeld in 1999 in their seminal paper "An algebraic method for public-key cryptography" [**2**]. The key exchange protocol proposed is particularly interesting as it is not an analog of a traditional cryptographic protocol, as many group theoretic protocols are. The authors proposed that the scheme be implemented in braid groups (see Chapter 7), which proved to be susceptible to heuristic attacks (see [**113**],[**163**],[**164**], and Chapter 5). Several years later Kahrobaei and Lam in [**117**] proposed Heisenberg groups as a platform group in the AAG protocol. Nonetheless, the notion proposed in [**2**] spurred the introduction of many other key exchange protocols using noncommutative groups.

Before introducing the scheme we would like to recall notation introduced in Chapter 1 that will be used throughout the text. Given two group elements g and a, the conjugation of g by a is denoted $g^a = a^{-1}ga$. The scheme is presented below.

AAG key exchange protocol: Let G be a finitely generated group with solvable word problem. Let S and T be two finitely generated subgroups of G, and let $\{s_1, \cdots, s_n\}$ and $\{t_1, \cdots, t_m\}$ be generating sets for S and T, respectively. Now suppose two people, Alice and Bob, want to agree on a key. The group G, its subgroups S and T and their generators are public information. Then

(a) Alice chooses a secret element $a \in S$ as a word in the generators of S, $a = s_{i_1}^{a_1} \cdots s_{i_l}^{a_l}$, and publishes $t_1^a, \ldots, t_m^a$.

(b) Bob chooses a secret element $b \in T$ as a word in the generators of T, $b = t_{j_1}^{b_1} \cdots t_{j_h}^{b_h}$, and publishes $s_1^b, \ldots, s_n^b$.

Based on this setup, Alice and Bob can use $[a, b]$ as a shared key. It is straightforward to observe that both Alice and Bob can compute $[a, b]$ readily, since

$$\begin{aligned} [a,b] &= (b^{-1})^a b = ((t_{j_1}^a)^{b_1} \cdots (t_{j_h}^a)^{b_h})^{-1} \cdot b && \text{computable for Bob} \\ &= a^{-1} a^b = a^{-1} \cdot (s_{i_1}^b)^{a_1} \cdots (s_{i_l}^b)^{a_l} && \text{computable for Alice.} \end{aligned}$$

For an adversary to obtain the shared key, it is sufficient to determine

- an element $a' \in S$ such that $t_j^a = t_j^{a'}$ in G for each $j = 1, \ldots m$,
- an element $b' \in T$ such that $s_i^b = s_i^{b'}$ in G for each $i = 1 \ldots n$.

Since $[a', b'] = [a, b]$ the adversary will determine the shared key (see [**163**]). Finding elements a' and b' requires the ability to solve the subgroup-restricted simultaneous conjugacy search problem, which is a variation of the simultaneous conjugacy search problem as noted in Chapter 2. The variation requires one to solve the simultaneous conjugacy search problem in a specific subgroup.

4.1.3. Ko, Lee, et al. key exchange. Ko, Lee, et al. introduced a noncommutative key exchange protocol in 2000 based on the Diffie-Hellman key exchange (see [**131**]). The authors proposed, as in the AAG protocol, that the key exchange protocol was implemented in braid groups. It has since been shown in ([**158**], [**68**]) that the protocol is not secure when implemented in the proposed platform groups as it is susceptible to length-based attacks (see Chapter 5). As in the AAG key exchange protocol conjugation of group elements is utilized; however, in this protocol commuting subgroups are required. Essentially, the key exchange protocol introduced by Ko, Lee, et al. is an analog of the Diffie-Hellman key exchange that uses conjugation instead of exponentiation. The scheme is presented below.

Ko, Lee, et al. key exchange: Let G be a noncommutative group with solvable word problem. Let $g \in G$ and let S and T be two commuting subgroups of G (i.e., for all $s \in S, t \in T, st = ts$). Suppose that two people, Alice and Bob, want to agree on a shared key. The group G, its element g and its subgroups S and T are public information. Then

a) Alice chooses a secret element $w \in S$ and publishes g^w.
b) Bob chooses a secret element $v \in T$ and publishes g^v.

In order to determine a shared key,

a) Alice can compute $(g^v)^w$.
b) Bob can compute $(g^w)^v$.

Since the subgroups S and T commute, we have that

$$(g^v)^w = (v^{-1}gv)^w = w^{-1}v^{-1}gvw = v^{-1}w^{-1}gwv = (w^{-1}gw)^v = (g^w)^v.$$

Hence, Alice and Bob can use $g^{wv} = g^{vw}$ as a shared secret.

The security of this protocol is based on the difficulty of the conjugacy search problem. Clearly, if an adversary can solve the conjugacy search problem then the shared key is easily computable. In [**195**], it was shown that in order for an adversary to obtain the shared key based on public information it is enough to solve the decomposition problem. The graph theoretic version of this problem was stated in Chapter 2. For convenience, we restate the problem in the context of groups.

Decomposition Problem: Given $w, x \in G$ and two subgroups $S, T \leq G$, find elements $s \in S, t \in T$ such that $x = swt$.

To see that solving this problem is sufficient to determine the shared key, suppose an adversary finds elements $v_1, v_2 \in T$ with $v_1 g v_2 = v^{-1}gv$ and elements

$w_1, w_2 \in S$ with $w_1 g w_2 = w^{-1} g w$. Then the adversary can compute

$$v_1 w_1 g w_2 v_2 = v_1 w^{-1} g w v_2 = w^{-1} v_1 g v_2 w = w^{-1} v^{-1} g v w,$$

which is the shared key, without ever solving the conjugacy search problem.

4.1.4. Noncommutative ElGamal. Kahrobaei and Khan proposed two noncommutative variations of the ElGamal encryption scheme in [**115**]. While the schemes can be used for general encryption, they are presented in the context of their application to a key exchange protocol. The authors propose that the scheme be implemented in polycyclic groups (see 10).

The first proposed scheme uses conjugation rather than exponentiation used in the original ElGamal scheme.

Noncommutative ElGamal key exchange using conjugacy: Let G be a finitely presented nonabelian group having solvable word problem. Let S and T be finitely generated proper commuting subgroups of G. Suppose that Alice would like to send the key x to Bob. Then the protocol can be implemented as follows.

(a) Bob chooses $s \in S, b \in G$ and publishes b and $c = b^s$ as his public keys, keeping s as his private key.

(b) Alice chooses a random $t \in T$, which she keeps secret, and sends

$$E = x^{(c^t)}$$

to Bob, along with the header

$$h = b^t.$$

(c) Bob then calculates $(b^t)^s = (b^s)^t = c^t$ using the header.

(d) He can now compute

$$E' = (c^t)^{-1},$$

which allows him to determine the key since

$$(x^{(c^t)})^{E'} = (x^{(c^t)})^{(c^t)^{-1}} = x.$$

The element $x \in G$ can now be used as a key.

Clearly if an adversary can solve the conjugacy search problem in the proposed platform, then he will obtain the key. Thus the security of this scheme rests on the assumption that there is no polynomial time algorithm for solving the conjugacy search problem for the group G.

4.1.4.1. *Noncommutative key exchange using power conjugacy.* The second proposed protocol considers the case where the conjugacy search problem is solvable in polynomial time. Kahrobaei and Khan modified the previous key exchange to address this possibility.

Noncommutative key exchange using power conjugacy: As in the previous protocol, G is a finitely presented nonabelian group with solvable word problem. Let S and T be finitely generated proper commuting subgroups of G. Suppose that Alice would like to send a key x to Bob. Then the protocol can be implemented as follows.

(a) Bob takes $s \in S$, $g \in G$, and $n \in \mathbb{N}$ and publishes $v = g^n$ and $w = g^s$ as his public keys. Bob keeps $n \in \mathbb{N}$ and $s \in S$ as his private keys. Note that v and w satisfy $w^n = v^s$.

(b) Alice chooses a random $m \in \mathbb{N}$ and $t \in T$. To encrypt x, Alice computes

$$E = x^{-1}t^{-1}(v)^m tx = x^{-1}t^{-1}g^{mn}tx$$

and sends it to Bob along with the header

$$h = t^{-1}w^m t = t^{-1}s^{-1}g^m st.$$

(c) Bob receives E and h, and computes

$$E' = sh^n s^{-1} = t^{-1}g^{mn}t.$$

Note that $E = x^{-1}E'x$, so if Bob can solve the conjugacy search problem, he can obtain $x \in G$, which can then serve as the common secret that can be used as a symmetric key for secure communication.

To recover Bob's private key from public information would require solving the equation $w^n = s^{-1}g^n s$ for n and s, given the public values g^n and w, which amounts to solving the power conjugacy search problem for G (see Chapter 2). Thus the security of this scheme rests on the assumption that there is no fast algorithm for solving the power conjugacy search problem for the group G.

4.1.5. Commuting action key exchange. In [**197**], Shpilrain and Zapata constructed a key exchange protocol, deemed commuting action key exchange (CAKE), and based on a generalization of the discrete log problem. The protocol is a generalization of the Diffie-Hellman key exchange protocol; thus, the Diffie-Hellman key exchange is an instance of CAKE.

Commuting action key exchange protocol: Let S and T be semigroups, $f : S \times T \to S$ a one-way action that is well defined for fixed values of T, and $\mathcal{H}$ is the set $\{A, B \subseteq T | \forall \alpha \in A, \forall \beta \in B, \alpha\beta = \beta\alpha\}$.

(a) The semigroup S, a word $w \in S$, and a generating set for each semigroup in $\mathcal{H}$ are made public.
(b) Alice chooses a secret word $\alpha \in A$ that satisfies $f(w, \alpha) \neq 1$ and sends $f(w, \alpha) = w\alpha$ to Bob.
(c) Bob chooses a secret word $\beta \in B$ that satisfies $f(w, \beta) \neq 1$ and sends $f(w, \beta) = w\beta$ to Alice.
(d) Alice computes $f(w\beta, \alpha) = w\beta\alpha$ and Bob computes $f(w\alpha, \beta) = w\alpha\beta$.
(e) Since $\alpha\beta = \beta\alpha$, the shared key is $w\beta\alpha = w\alpha\beta$.

In [**197**], Shpilrain and Zapata propose that CAKE be implemented in small cancellation groups and conjugation as the group action. See Chapter 9 for more details on small cancellation groups and their use in CAKE.

4.1.6. Key exchange protocols based on semidirect product. Here we present two key exchange protocols based on a more complex group structure, the semidirect product of two groups (or semigroups), which was defined in the first chapter.

An interesting case of the semidirect product construction in our context is where the group H is a subgroup of the group $Aut(G)$. If $H = Aut(G)$, then the corresponding semidirect product is called the *holomorph* of the group G, and is denoted $Hol(G)$. $Hol(G)$ is the set of all pairs (g, ϕ), where $g \in G$, $\phi \in Aut(G)$, with the group operation given by $(g, \phi) \cdot (g', \phi') = (\phi'(g) \cdot g', \phi \cdot \phi')$.

4.1.6.1. *HKKS key exchange protocol.* In [**96**], Habeeb, Kahrobaei, Koupparis, and Shpilrain (HKKS) proposed a key exchange protocol utilizing the construction of the semidirect product where H was a subgroup of $Aut(G)$. The simplest implementation of the protocol utilizes a cyclic subgroup (or a cyclic subsemigroup) of the group $Aut(G)$ (respectively, of the semigroup $End(G)$ of endomorphisms) instead of the entire group of automorphisms of G.

Let G be a group (or semigroup). An element $g \in G$ is chosen and made public as well as an arbitrary automorphism $\phi \in Aut(G)$ (or an arbitrary endomorphism $\phi \in End(G)$). Bob chooses a private $n \in \mathbb{N}$, while Alice chooses a private $m \in \mathbb{N}$. An element of the semidirect product is of the form (g, ϕ^r), where $g \in G$, $r \in \mathbb{N}$, and the product of two elements of this form is: $(g, \phi^r) \cdot (h, \phi^s) = (\phi^s(g) \cdot h,\ \phi^{r+s})$.

(a) Alice computes $(g, \phi)^m = (\phi^{m-1}(g) \cdots \phi^2(g) \cdot \phi(g) \cdot g,\ \phi^m)$ and only sends the first component of this pair to Bob. Thus, she only sends to Bob the element $a = \phi^{m-1}(g) \cdots \phi^2(g) \cdot \phi(g) \cdot g$ of the group (or semigroup) G.

(b) Bob computes $(g, \phi)^n = (\phi^{n-1}(g) \cdots \phi^2(g) \cdot \phi(g) \cdot g,\ \phi^n)$ and only sends the first component of this pair to Alice. Thus, he only sends to Alice the element $b = \phi^{n-1}(g) \cdots \phi^2(g) \cdot \phi(g) \cdot g$ of the group (or semigroup) G.

(c) Alice computes $(b, x) \cdot (a,\ \phi^m) = (\phi^m(b) \cdot a,\ x \cdot \phi^m)$. Her key is now $K_A = \phi^m(b) \cdot a$.

(d) Bob computes $(a, y) \cdot (b,\ \phi^n) = (\phi^n(a) \cdot b,\ y \cdot \phi^n)$. His key is now $K_B = \phi^n(a) \cdot b$.

(e) Since $(b, x) \cdot (a,\ \phi^m) = (a,\ y) \cdot (b,\ \phi^n) = (g,\ \phi)^{m+n}$, we have $K_A = K_B = K$, the shared secret key.

It should be noted that in step (c) of the protocol, Alice cannot compute the second component, $x \cdot \phi^m$, in the product. Similarly, in step (d) of the protocol Bob cannot compute the second component, $y \cdot \phi^n$, in the product since $x = \phi^n$ and $y = \phi^m$ are unknown to Alice and Bob, respectively.

The protocol above can be implemented in any noncommutative group or semigroup as long as ϕ is a nontrivial inner automorphism, which is an automorphism arising from conjugation. The authors in [**96**] proposed the protocol be implemented in matrices over group rings of a symmetric group. In the proposed platform, the protocol was susceptible to linear algebra attacks (see Chapter 5, [**161**] and [**188**]). In [**120**], free nilpotent p-groups were proposed as the platform for the above mentioned protocol.

4.1.6.2. *A matrix action key exchange (MAKE).* Rahman and Shpilrain proposed a public-key exchange protocol in [**180**] based on the semidirect product of two cyclic groups (or semigroups of matrices over $\mathbb{Z}_p$) where one group (or semigroup) is additive and the other multiplicative. Here we follow the exposition of [**180**].

For $n \in \mathbb{N}$ and p prime, let G be the additive group of $n \times n$ matrices over $\mathbb{Z}_p$, denoted $M_n(\mathbb{Z}_p)$, and $S = \{(H_1^i, H_2^i) : i \in \mathbb{N}\}$ be the semigroup generated by noninvertible matrices $H_1, H_2 \in M_n(\mathbb{Z}_p)$. The action of S on G is defined by $(H_1^i, H_2^i) \cdot M = H_1^i M H_2^i$. Note that S must be commutative to satisfy the axioms of an action, which is why a cyclic group (or semigroup) is used. This action induces a homomorphism into the automorphism group of G, allowing one to define the semidirect product of G by S with multiplication

$$(M, (H_1^i, H_2^i))(M', (H_1^j, H_2^j)) = (H_1^j M H_2^j + M', (H_1^{i+j}, H_2^{i+j}))$$

For any choice of H_1, H_2, exponentiation has the form

$$(M, (H_1, H_2))^n = \left(\sum_{i=0}^{n-1} H_1^i M H_2^i, (H_1^n, H_2^n) \right).$$

The key exchange protocol is as follows. Suppose Alice and Bob wish to agree on a shared, private key by communicating over an insecure channel with public data M, H_1, H_2 available.

(a) Alice picks random $x \in \mathbb{N}$ and calculates $(M, (H_1, H_2))^x = (A, (H_1^x, H_2^x))$ and sends A to Bob.

(b) Bob similarly calculates a value B corresponding to random $y \in \mathbb{N}$, and sends it to Alice.

(c) Alice calculates $(B, *)(A, (H_1^x, H_2^x)) = (H_1^x B H_2^x + A, **)$ and arrives at her key $K_A = H_1^x B H_2^x + A$. She does not actually calculate the product explicitly since she does not know the value of $*$; however, it is not required to calculate the first component of the product.

(d) Bob similarly calculates his key as $K_B = H_1^y A H_2^y + B$.

Since $A = \sum_{i=0}^{x-1} H_1^i M H_2^i$, $B = \sum_{i=0}^{y-1} H_1^i M H_2^i$, we have

$$\begin{aligned} H_1^x B H_2^x + A &= H_1^x \left(\sum_{i=0}^{y-1} H_1^i M H_2^i \right) H_2^x + A \\ &= \sum_{i=x}^{x+y-1} H_1^i M H_2^i + \sum_{i=0}^{x-1} H_1^i M H_2^i \\ &= \sum_{i=y}^{x+y-1} H_1^i M H_2^i + \sum_{i=0}^{y-1} H_1^i M H_2^i \\ &= H_1^y A H_2^y + B. \end{aligned}$$

Alice and Bob therefore both arrive at the same shared key $K = K_A = K_B$.

This protocol was also found to be vulnerable to linear algebra attacks like the HKKS scheme. This protocol was also found to be vulnerable to another attack [**22**] that relies on the commutativity of the underlying ring. In [**10**], it was shown that under certain circumstances using a noncommutative underlying ring will have the same vulnerability. We refer the reader to Chapter 5 for more details.

4.2. Noncommutative digital signatures

4.2.1. Introduction. Several noncommutative digital signatures have been proposed over the last twenty years after the introduction of braid group cryptography. Some of these schemes were based on the difficulty of the conjugacy search problem and are now known to be susceptible to length-based attacks. Hence, braid groups are not suitable platform groups for these schemes. More recent digital signature schemes have suggested the use of noncommutative division semirings and noncommutative groups over matrices with elements from a finite field. The main references for this section are [**116**] and [**166**].

4.2.2. Kahrobaei-Kaoupparis digital signature scheme. In [**116**], Kahrobaei and Koupparis proposed a noncommutative digital signature scheme inspired by Schnorr's digital signature scheme for commutative groups, which uses string concatenation and hash functions. As the scheme must be implemented in a platform group where the conjugacy search problem is hard, Kahrobaei and Koupparis proposed using polycyclic groups (see Chapter 10) or any group deemed secure against length-based attacks.

Before introducing the scheme we would like to clarify some notation and give some necessary definitions. Given two bit strings n and m, $n||m$ is the concatenation of n and m.

DEFINITION 4.1. *A hash function is a function that takes an input of arbitrary length and returns an output with a fixed length.*

DEFINITION 4.2. *A hash function H is said to be collision free if for two distinct inputs x and y $H(x) \neq H(y)$*

Due to the fact that a hash function takes inputs of arbitrary lengths to outputs of the same length, hash functions are not typically collision-free. Rather than being collision-free, it suffices that a hash function is collision-resistant. A hash function, H is collision-resistant if there is no efficient algorithm to find x and y such that $x \neq y$ and $H(x) = H(y)$ The scheme is presented below.

DEFINITION 4.3. *Given a finitely generated group G, the growth rate is determined by its* growth function $\gamma : \mathbb{N} \longrightarrow \mathbb{R}$ *defined as* $\gamma(n) = \#\{w \in G : l(w) \leq n\}$, *where $l(w)$ is the length of w as a word in the generators of G.*

Kahrobaei-Koupparis digital signature:

Let G be an infinite finitely presented group with an exponential growth rate. In addition, the conjugacy search problem should be intractable in G. Let $f : G \to \{0,1\}^*$ be a mapping that sends the group to a binary representation that can be digitally encoded. Let H be a collision-free hash function mapping into G.

Public-key generation:

- To begin, the signer, Alice, chooses a group element $g \in G$ with trivial centralizer, a private key $s \in G$, and an integer $n \in \mathbb{N}$. The natural number n chosen should be a highly composite number; that is, the number of divisors for n should be greater than that of any smaller positive integer.
- Alice then computes the value $x = s^{-1}g^n = g^{ns}$, and publishes x.
- Alice picks an element $t \in G$, a random factorization of $n = n_i n_j$ and computes the key $y = t^{-1}g^{n_i}t = g^{n_i t}$.

Signing algorithm:

- To sign the message m, Alice computes the following.

$$h = H(m||f(y))$$
$$\alpha = t^{-1}shy$$

- Alice publishes the triple (y, α, n_j) as her signature along with the message m.

Verification algorithm:

- To verify the signature compute $h' = H(m||f(y))$.
- The signature is valid if and only if

$$y^{n_j \alpha} = x^{h' y}.$$

In order to see why the verification algorithm works, consider a signature (y, α, n_j) generated by Alice, message m and the public key x. Bob computes $h' = H(m||f(y))$ and verifies that

$$y^{n_j \alpha} = x^{h' y}$$

holds. Note that the left-hand side yields

$$y^{n_j \alpha} = y^{n_j t^{-1} shy} = g^{n_i t n_j t^{-1} shy} = g^{n_j n_i t t^{-1} shy} = g^{nshy} = x^{hy}.$$

If the signature is valid, $h = h'$ and the equality will hold. Otherwise, the signature will be invalid.

4.2.3. Mohan and Anjaneyulu digital signature scheme. In [**166**], Mohan and Anjaneyulu proposed a digital signature scheme that connects the conjugacy problem and the discrete log problem on a noncommutative group of matrices over a finite field. The scheme is presented below.

Mohan and Anjaneyulu digital signature scheme: Let G be a noncommutative group generated over a field of elements F where K_1 and K_2 are commuting subgroups of G. G, K_1, K_2, an element $X \in G$ and an integer n are public information.

Key generation: Alice arbitrarily chooses a secret integer $a \in \{1, 2, \ldots, n-1\}$, a secret matrix $U \in K_1$ with $UX \neq XU$ and $V \in K_2$. Alice computes the public key $\alpha = U^{-1} X^a U$.

Signature generation: Let $G = GL_2(\mathbb{F}_q)$ and $X \in GL_2(\mathbb{F}_q)$ be of order n. In order to sign the message M Alice performs the following computations:

$$\beta = V^{-1} X^a U,$$
$$\gamma = V^{-1} X^a V,$$
$$\delta = U^{-1} X^a V.$$

Alice's signature is then the tuple $(\beta, \gamma, \delta, M)$, which she sends to Bob for verification.

Verification algorithm: Upon receiving the signature $(\beta, \gamma, \delta, M)$ from Alice, Bob begins by computing $\theta = \alpha\beta^{-1}$. If $\delta = \alpha\beta^{-1}$, Bob accepts the signature; otherwise, he rejects the signature.

To see that Bob will always accept Alice's signature as long as she follows the appropriate procedure, recall that $\alpha = U^{-1} X^a U$ is Alice's public key, $\theta = \alpha\beta^{-1}$ and $(\beta, \gamma, \delta, M)$ is the signature. In order to verify the signature the equality $\delta = \theta\gamma$

must hold. Recall that

$$\beta = V^{-1}X^aU,$$
$$\gamma = V^{-1}X^aV,$$
$$\delta = U^{-1}X^aV.$$

Hence,

$$\begin{aligned}\theta\gamma &= (U^{-1}X^aU)(V^{-1}X^aU)^{-1}\gamma \\ &= (U^{-1}X^aU)(U^{-1}(X^a)^{-1}V)\gamma \\ &= U^{-1}V\gamma \\ &= (U^{-1}V)(V^{-1}X^aV) \\ &= U^{-1}X^aV \\ &= \delta.\end{aligned}$$

Therefore, when Alice follows the signing algorithm appropriately Bob always accepts the signature.

For the interested reader we refer [**166**] for a discussion on the security of the scheme against existential forgery, signature repudiation, selective forgery, and a total break of the scheme.

4.3. Noncommutative secret sharing schemes

4.3.1. Introduction. Many noncommutative cryptographic protocols are motivated by their commutative counterparts. This holds true for noncommutative secret sharing schemes as we will introduce schemes based on Shamir's secret sharing scheme. We begin by discussing the Panagopoulos threshold scheme as it has motivated the construction of some noncommutative threshold schemes. These include the Habeeb-Kahrobaei-Shpilrain (HKS) secret sharing scheme, which is a hybrid of Shamir's secret sharing scheme, essentially combining an (n, n)-threshold scheme with Shamir's secret sharing scheme, that provides some advantages to Shamir's scheme, the Cavallo and Kahrobaei secret sharing scheme and threshold schemes based on Nielsen transformations. Throughout this chapter we will refer to [**175**], [**27**] and [**59**].

4.3.2. Panagopoulos's threshold scheme. In [**175**], D. Panagopoulos suggested a (k, n)-threshold scheme based on the word problem in which the secret is a binary sequence. The word problem in the platform group should be efficiently solvable for the feasibility of the scheme. The scheme can be summarized as follows.

(a) Choose a group with solvable word problem and finite presentation, $\langle x_1, x_2, \ldots, x_s | r_1, r_2, \ldots, r_m\rangle$, where $m = \binom{n-1}{k}$. The generators, $x_1, x_2, \ldots, x_s$, are public information.

The dealer begins by

(b) Fixing a sequence $A_1, \ldots, A_m$ of subsets of $\{1, \ldots, n\}$ with $k-1$ elements.

(c) The subsets $R_1, \ldots, R_n$ of $\{r_1, \ldots, r_m\}$ are defined using the property

$$a_j \in R_i \text{ if and only if } i \notin A_j \text{ for } j = 1, \ldots, m \text{ and } i = 1, \ldots n.$$

The dealer then distributes the set R_i to the ith participant. Each set is the participant's share.

Let the secret be the binary sequence $a_1 \cdots a_l$. The dealer then constructs and distributes words, $w_1, \ldots, w_l$, in G in such a way that $w_i =_G 1$ if and only if $a_i = 1$ for $i = 1, 2, \ldots, l$.

In order to recover the secret, the participants take the union of their sets R_i. Any k participants can recover the secret, as the union of the sets is the set of relators if and only if at least k participants contribute their shares. One can see this by noting that by definition of the sets R_i a relator r_j is not contained in exactly $k-1$ of the subsets $R_1, \ldots, R_n$ for $j = 1, \ldots, m$. Hence, it follows that r_j is contained in any union of k of the sets $R_1, \ldots, R_n$; however, if the union of any $k-1$ of them is taken there exists a j with r_j not in the union. Since the set of all the relators is required to determine the secret, the participants do not gain any knowledge about the secret unless at least k of them pool their shares. In order to ensure security each word w_i must be chosen so that most of the relators are used whenever $w_1 = 1$.

4.3.3. A hybrid secret sharing scheme. As mentioned above the HKS secret sharing scheme is a hybrid scheme, essentially combining an (n, n)-threshold scheme with Shamir's secret sharing scheme. The objective of the (n, n)-threshold scheme is to to distribute a k-column vector, the secret, $C = \begin{pmatrix} c_1 \\ c_2 \\ \vdots \\ c_k \end{pmatrix}$ consisting of zeros and ones among n participants in such a way that the vector can be retrieved only when all n participants cooperate. A set of group generators $X = \{x_1, \ldots, x_m\}$ is made public. The secret can be recovered by implementing the following scheme:

(a) The dealer distributes over a secure channel to each participant P_j a set of words R_j in the alphabet $X^{\pm 1} = \{x_1^{\pm 1}, \ldots, x_m^{\pm 1}\}$ such that each group $G_j = \langle x_1, \ldots, x_m \mid R_j \rangle$ has efficiently solvable word problem.

(b) The dealer then distributes words $w_{1j}, \ldots, w_{kj}$ in the generators $x_1, \ldots, x_m$ over an open channel to each participant P_j, $1 \le j \le n$. The words are chosen so that the number of words w_{ij} such that $w_{ij} = 1$ in G_j is even if $c_i = 0$ and odd if $c_i = 1$.

(c) Participant P_j then checks, for each i, whether the word $w_i = 1$ in G_j or not. After that, each participant P_j can make a column of 0's and 1's, $C_j = \begin{pmatrix} c_{1j} \\ c_{2j} \\ \vdots \\ c_{kj} \end{pmatrix}$, by setting $c_{ij} = 1$ if $w_i = 1$ in G_j and 0 otherwise.

(d) The participants then construct the secret by forming the column vector $C = \sum_{j=1}^{n} C_j$, where the sum of the entries is taken modulo 2.

The HKS scheme implements Shamir's scheme with the idea of the scheme above. As in Shamir's scheme, the secret is an element $x \in \mathbb{Z}_p$ and the dealer chooses a polynomial f of degree $t-1$ such that $f(0) = x$. In addition the dealer determines integers $y_i = f(i)$ that need to be distributed to participant P_i for $1 \le i \le n$. A set of generators $\{x_1, \cdots, x_m\}$ is made public. The scheme is as follows.

(a) The dealer distributes over a secure channel to each participant P_j a set of relators R_j such that each group $G_j = \langle x_1, \cdots, x_m | R_j \rangle$ has efficiently solvable word problem.

(b) The dealer then distributes over open channels k-column vectors $b_j = \begin{pmatrix} b_{1j} \\ b_{2j} \\ \vdots \\ b_{kj} \end{pmatrix}$ for $1 \leq j \leq n$ of words in $x_1, \cdots, x_m$ to each participant. The column vectors b_j are chosen so that in participant P_j's group precisely y_j of the words are equal to 1.

(c) Participant j then checks if each word $b_{ij} = 1$ in his or her group G_j. The total number of words $b_{ij} = 1$ in G_j is the number y_j.

(d) Each participant now has a point $f(i) = y_i$ of the polynomial. Using polynomial interpolation any t participants can now recover the polynomial f, and hence the secret $f(0)$.

This scheme has clear advantages over Shamir's scheme. The first is that the secret need not be distributed over secure channels. Secondly, the shares (i.e., relators) do not depend on the secret. This allows for the same shares to be used for various secrets. One should note that while this is an advantage of the HKS scheme, it poses a security concern for the scheme. Over time, as more secrets are shared, the set of relators distributed to each participant becomes less secure.

In order for the scheme to be practical, the word problem must be efficiently solvable for each group presentation. For this reason, Habeeb, Kahrobaei, and Shplirain proposed small cancellation groups satisfying $C'\left(\frac{1}{6}\right)$, since the word problem is solvable in linear time using a variation of Dehn's algorithm. In addition, a generic finitely presented group is a small cancellation group. Hence, a randomly selected set of relators will define a small cancellation group with overwhelming probability. This allows for a straightforward method to generate the secret shares for each participant. For more on small cancellation groups, see Chapter 9.

4.3.4. Secret sharing scheme using shortlex order. The next secret sharing scheme we will discuss, due to Cavallo and Kahrobaei, is a variation on the HKS scheme based on the shortlex order on a free group (see [**27**]). We begin by defining a shortlex ordering on a free group.

DEFINITION 4.4. *Let $X = \{x_1, x_2, \ldots, x_n\}$ and $G = \langle X \rangle$. A* shortlex ordering *on G is induced by an ordering on $X^{\pm 1}$ and can be defined as follows. Given distinct reduced words $w = x_{i_1} \cdots x_{i_p}$ and $u = x_{j_1} \cdots x_{j_k}$ we say $w < m$ if and only if either*

- *$|w| < |u|$ or*
- *if $p = k$ and $x_{i_m} < x_{j_m}$ where $m = \min_\alpha \{x_{i_\alpha} \neq x_{j_\alpha}\}$.*

Cavallo and Kahrobaei used the shortlex ordering to modify the HKS (k, n)-threshold secret sharing scheme as follows:

(a) The dealer begins by publishing the letters X. Over a private channel the dealer sends a set of words R_i in $X^{\pm 1}$ to each participant, P_i, such that $G_i = \langle X | R_i \rangle$ is a group with an efficient algorithm to reduce words with respect to R_i or an efficient algorithm to compute normal forms.

(b) The dealer chooses a secret $s \in \mathbb{Z}_p$, where p is large prime greater than n and generates a random polynomial $f \in \mathbb{Z}_p[x]$ with $f(0) = s$.

(c) The dealer assigns each participant a public element $n_i \in \mathbb{Z}_p$. The dealer then computes $f(n_i)$ and finds $s_i \in F(X)$ such that s_i is the $f(n_i)^{th}$ word in $F(X)$.

(d) The dealer publishes a word w_i that reduces to s_i in G_i.

(e) Each participant P_i reduces w_i to s_i and then computes its position in $F(X)$ under the shortlex ordering.
(f) The participants can then use their shares to obtain the secret using polynomial interpolation.

Since this scheme only requires each participant to be able to efficiently reduce words, it is not necessary to have efficiently solvable word problem. In some cases, the word problem is more difficult than reducing the word relative to a set of relators. The main advantage of this scheme over the HKS scheme is that the participants are only required to reduce one word rather than several words corresponding to the length of the secret.

REMARK 4.5. For efficiency, when the participants reduce each w_i to s_i, conjugated products of relators can be interspersed between the letters of s_i.

REMARK 4.6. For the feasibility of the scheme, the reduction algorithm used in each group G_i must reduce each w_i uniquely to s_i.

4.3.5. Secret sharing based on Nielsen transformations. In [**59**], Fine, Moldenhauer, and Rosenberger proposed two modifications of Panagopoulos's threshold scheme utilizing regular Nielsen transformations. For clarity, we provide the reader with the relevant definitions before introducing the scheme. We follow the exposition of [**59**].

DEFINITION 4.7. *Let F be the free group on the generating set $\{x_1, x_2, \ldots, x_m\}$ with $m \geq 2$ and let U be the subset of reduced words, $U = \{u_1, u_2, \ldots, u_n\} \subset F$ with $n \geq 2$. An* elementary Nielsen transformation *on U is one of the following transformations:*

- *(T1): Replace some u_i by u_1^{-1}.*
- *(T2): Replace some u_i by $u_i u_j$ where $j \neq i$.*
- *(T3): Remove some u_i where $u_i = 1$.*

DEFINITION 4.8. *A finite product of elementary Nielsen transformations is called a* Nielsen transformation.

DEFINITION 4.9. *A Nielsen transformation is called* regular *if it is a finite product of transformations of types (T1) and (T2).*

DEFINITION 4.10. *A finite set U in F is called* Nielsen-reduced *if for any three elements $t_1, t_2, t_3 \in U^{\pm}$ the following conditions hold.*
(a) $t_1 \neq 1$,
(b) *$t_1 t_2 \neq 1$ implies that $t_1 t_2 > |t_1|$ and $t_1 t_2 > |t_2|$,*
(c) *$t_1 t_2 \neq 1$ and $t_2 t_3 \neq 1$ imply that $|t_1 t_2 t_3| > |t_1| - |t_2| + |t_3|$,*

where $|t_i|$ is the free length of t_i.

In the scheme presented by Fine et al., the secret is a rational number rather than a binary sequence as in Panagopoulos's threshold scheme. As in the original scheme $m = \binom{n-1}{k}$. The scheme is as follows.

Let $x_1, \ldots, x_m$ be generators on the free group $F(x_1, x_2, \ldots, x_m)$. Let $N_{q_i} \in SL(2, \mathbb{Q})$ be matrices of the form

$$\begin{pmatrix} -q_i & -1 + q_i^2 \\ 1 & -q_i \end{pmatrix},$$

satisfying $q_1 \geq 2$ and $q_{i+1} \geq q_i + 3$. The secret is then a rational number

$$\sum_{i=1}^{m} \frac{1}{|tr(N_{q_i})|} = \sum_{i=1}^{m} \frac{1}{2|q_i|}.$$

The dealer implements share distribution as follows.

(a) Begin with a set of pairs $(x_1, N_{q_1}), \ldots, (x_m, N_{q_m})$.

(b) Apply a set of regular Nielsen transformations to the set of pairs in (a) to obtain a new set $(v_1, M_1), \ldots, (v_m, M_m)$.

(c) Distribute subsets as in Panagopoulos's scheme.

As in Panagopoulous's scheme in order to recover the secret condition $C'(1/6)$, the participants must take a union of their shares. If k participants gather, then they have the capability to recover the set $(v_1, M_1), \ldots, (v_m, M_m)$. The participants may then apply a sequence of Nielsen transformations to Nielsen-reduce the first components to $x_1, \ldots, x_m$. This reduction will in turn allow the participants to obtain the original matrices $N_{q_1}, \ldots, N_{q_m}$. With this knowledge the participants can compute the secret, $\sum_{i=1}^{m} \frac{1}{|tr(N_{q_i})|}$.

The second modification of Panagopoulos's scheme introduced by Fine, Moldenhauer, and Rosenberger can be described as follows.

(a) The dealer begins with a Nielsen reduced set of words $\overline{u} = (u_1, u_2, \ldots, u_m)$ and the secret is the rational number $\sum_{i=1}^{m} \frac{1}{|u_i|}$.

(b) The dealer then applies a sequence of regular Nielsen transformations to $\overline{u}$ to obtain the set $\overline{v} = (v_1, v_2, \ldots, v_m)$.

(c) The dealer then distributes subsets of $\overline{v}$ to the participants as in Panagopoulos's scheme.

In order to recover the secret, t participants take the union of their shares to obtain the set $\overline{v}$. The participants then Nielsen-reduce the set $\overline{v}$ to obtain a set $\overline{u'} = (u_1', u_2, \ldots, u_m')$. Although the sets $\overline{u}$ and $\overline{u}'$ are different, the sums $\sum_{i=1}^{m} \frac{1}{|u_i|}$ and $\sum_{i=1}^{m} \frac{1}{|u_i'|}$ are equal. Hence, they can recover the secret.

The secret sharing schemes proposed by Fine et al. provide an interesting use of Nielsen transformations in cryptography. In fact, the authors based cryptosystems on these schemes. We refer the interested reader to the original paper [**59**].

CHAPTER 5

Attacks

5.1. Introduction

As noncommutative cryptographic protocols arise, it becomes necessary to develop the cryptanalysis of such protocols. In fact, both topics should be considered of equal importance, as the creation of a protocol whose security is unknown is of no practical use. To illustrate the idea of cryptanalysis we begin by introducing the notion of a brute force attack, which is the equivalent of trying every possible key until the right one is found. For example, if an adversary were to implement a brute force attack on the Diffie-Hellman key exchange he would proceed as follows. The adversary has the public information: a large prime p, primitive root g, $X = g^x \mod p$ and $Y = g^y \mod p$. From this information, the adversary needs to compute the shared key, $g^{xy} \mod p$. In order to do this, the adversary would compute

$$\begin{aligned} g^1 &\mod p \\ g^2 &\mod p \\ &\vdots \end{aligned}$$

until one of the values matches either X or Y. From this, the adversary obtains the secret key (either x or y) and can compute the shared key $k = g^{xy} \mod p$.

Next we will review several known attacks on the protocols discussed in Chapter 4. This chapter is organized as follows. We begin with the length-based attack (see [**113**] and [**163**]), which has been proven effective in breaking the AAG protocol in the originally proposed braid groups. We will then introduce the linear centralizer attack (see [**196**] and [**205**]), which also breaks the AAG protocol. Next, the linear decomposition attack (see [**161**]) is introduced, which attacks several group-based cryptosystems. We conclude the chapter with the quotient attack (see [**125**] and [**164**]), which solves decision problems in groups.

5.2. Length-based attack

Hughes and Tannenbaum initially proposed length-based attack, in [**113**], on the Anshel-Anshel-Goldfeld protocol. The length-based attack (LBA) is a heuristic descent method for solving the subgroup restricted simultaneous conjugacy search problem, in which the security of the AAG protocol is based on.

Here we follow the exposition of [**163**]. The length-based attack determines Alice's (or Bob's) key by considering conjugating by the element a as a sequence of conjugations. Recall, in the AAG protocol Alice chooses a secret word $a = s_{i_1}^{a_1} s_{i_2}^{a_2} \dots s_{i_l}^{a_l}$ and then sends Bobs $b_i^a = a^{-1}b_i a$. One can think of the computation $a^{-1}b_i a$ as a sequence of conjugations beginning with $s_{i_1}^{-a_1} a s_{i_1}^{a_1}$, then conjugating by $s_{i_2}^{a_2}$, followed by $s_{i_3}^{a_3}$ and continuing in this manner until $s_{i_l}^{a_l}$. The LBA begins with the last step in the conjugating sequence, and recovers a conjugating factor in each step of the sequence.

In order to implement the length-based attack, a length function l must be utilized with the property that $l(a^{-1}ba) > l(b)$ for the majority of elements $a, b \in G$. In [**163**], the approximation function for geodesic length was used and three variations of the LBA were presented. We begin with LBA with backtracking, which enumerates all possible sequences of conjugations that decrease the length of a tuple.

LBA with backtracking

Input: Tuples $\overline{s} = (s_1, \ldots, s_n)$, $\overline{t} = (b_1, \ldots, b_m)$ and $\overline{b} = (b_1^a, \ldots, b_m^a)$ where a is a word in $s_1, \ldots, s_n$. Output: A word x in $s_1, \ldots, s_n$ such that $\overline{t}^x = \overline{b}$ or FAIL.

(a) Initialize a set $S = \{(\overline{b}, e)\}$ where e is the identity of the group G.
(b) If $S = \emptyset$, then output FAIL.
(c) Choose a pair $(\overline{c}, x) \in S$ with minimal $l(c)$. Remove $(\overline{c}, x)$ from S.
(d) For each $i = 1, \ldots, m$, $\epsilon = \pm 1$, compute $\delta_{i,\epsilon} = l(\overline{c}) - l(\overline{c}^{s_i^{\epsilon}})$.
(e) If $\delta_{i,\epsilon} > 0$, then add $(\overline{c}^{s_i^{\epsilon}}, x s_i^{\epsilon})$ into S.
(f) If $\overline{c}^{s_i^{\epsilon}} = \overline{s}$, then output $x s_i^{\epsilon}$.
(g) Otherwise go to (2).

Next we will introduce best descend LBA, which is a length-based attack where in each step the conjugator that gives the maximal decrease among length is chosen. The algorithm is the same as the LBA with backtracking except for the modification of step (e) to only add the tuple corresponding to the maximal positive $\delta_{i,\epsilon}$ to S.

Best descend LBA

Input: Tuples $\overline{s} = (s_1, \ldots, s_n)$, $\overline{t} = (b_1, \ldots, b_m)$ and $\overline{b} = (b_1^a, \ldots, b_m^a)$, where a is a word in $s_1, \ldots, s_n$. Output: A word x in $s_1, \ldots, s_n$ such that $\overline{t}^x = \overline{b}$ or FAIL.

(a) Initialize a set $S = \{(\overline{b}, e)\}$ where e is the identity of the group G.
(b) If $S = \emptyset$, then output FAIL.
(c) Choose a pair $(\overline{c}, x) \in S$ with minimal $l(c)$. Remove $(\overline{c}, x)$ from S.
(d) For each $i = 1, \ldots, m$, $\epsilon = \pm 1$, compute $\delta_{i,\epsilon} = l(\overline{c}) - l(\overline{c}^{s_i^{\epsilon}})$.
(e) Choose the greatest $\delta_{i,\epsilon} > 0$ (if it exists) and then add $(\overline{c}^{s_i^{\epsilon}}, x s_i^{\epsilon})$ into S.
(f) If $\overline{c}^{s_i^{\epsilon}} = \overline{s}$, then output $x s_i^{\epsilon}$.
(g) Otherwise, go to (2).

The final version of the length-based attack is the generalized LBA, which is an LBA with backtracking where the set of elements in Bob's (respectively Alice's) public sets are extended.

Generalized LBA

Input: Tuples $\overline{s} = (s_1, \ldots, s_n)$, $\overline{t} = (b_1, \ldots, b_m)$ and $\overline{b} = (b_1^a, \ldots, b_m^a)$ where a is a word in $s_1, \ldots, s_n$. Output: A word x in $s_1, \ldots, s_n$ such that $\overline{t}^x = \overline{b}$ or FAIL.

(a) Extend $\overline{s}$ with products $s_{j_1}^{\epsilon_1} \cdots a_{j_k}^{\epsilon_k}$ where k is limited by a contstant.
(b) Run LBA with backtracking or best descend LBA with the new tuples.

5.3. Linear centralizer attack

The linear centralizer method, introduced first in [**196**] and later developed in [**205**], is a method to solve the commutator key exchange problem (CKEP) and can be applied in many areas of group-based cryptography. The CKEP can be stated as follows.
Commutator Key Exchange Problem: Let G be a group with $a_1, \dots, a_k, b_1, \dots, b_k \in G$. Let $a \in \langle a_1, \dots, a_k \rangle$ and $b \in \langle b_1, \dots, b_k \rangle$. Given the elements

$$a_1, \dots, a_k, b_1, \dots, b_k, a_1^{b_1}, \dots, a_k^{b_k}, b_1^{a_1}, \dots, b_k^{a_k},$$

compute $[a, b]$.

Recall that the AAG protocol (see Chapter 4), also referred to as the commutator key exchange, is based on the subgroup-restricted simultaneous conjugacy search problem. The public information in the AAG protocol is a group G, subgroups S and T with generators $s_1, \dots, s_n$ and $t_1, \dots, t_m$ and the elements

$$t_1^a, \dots, t_m^a \text{ and } s_1^b, \dots, s_n^b.$$

The words a and b are elements in S and T, respectively; these are secret. The shared key is the commutator $[a, b]$. It is clear that solving the CKEP would determine the shared key in the AAG key exchange. Since the linear centralizer method solves the CKEP, this would result in a successful attack on the AAG protocol.

Now we present the linear centralized attack applied to matrix groups (see Chapter 1). This attack can be applied to other classes of groups including braid groups and Thompson groups.

Let $M_n(\mathbb{F})$ be the group of matrices $n \times n$ with entries in the field $\mathbb{F}$. Recall that for a subset $X \subseteq M_n(F)$, the *centralizer* of X in $M_n(\mathbb{F})$ is the set

$$C(X) = \{g \in M_n(\mathbb{F}) : gx = xg \text{ for all } x \in X\}.$$

Sometimes we write, for $a_1, \dots, a_k \in M_n(\mathbb{F})$, $C(\{a_1, \dots, a_k\}) = C(a_1, \dots, a_k)$. The algorithm works in two phases (offline and online) as follows:

Offline phase.

Input: $b_1, \dots, b_k \in G$.

Output: A basis for $C(S)$.

Algorithm: First compute a basis $S = \{s_1, \dots s_d\}$ for $C(b_1, \dots, b_k)$, by solving the following:

$$\begin{cases} b_1 \cdot x = x \cdot b_1, \\ \vdots \\ b_k \cdot x = x \cdot b_k. \end{cases}$$

Then, compute a basis for $C(S) = C(C(b_1, \dots, b_k))$ in the following way:

$$\begin{cases} s_1 \cdot x = x \cdot s_1, \\ \vdots \\ s_d \cdot x = x \cdot s_d. \end{cases}$$

Online phase. Fix $a \in \langle a_1, \ldots, a_k \rangle$ and $b \in \langle b_1, \ldots, b_k \rangle$, where a_i and b_i are elements of G for $1 \leq i \leq k$.
Input: $a_1, \ldots, a_k, b_1, \ldots, b_k, a_1^b, \ldots, a_k^b, b_1^a, \ldots, b_k^a$ where a and b are unknown.
Output: $[x, y]$.
Algorithm: Solve the following system:

$$\begin{cases} b_1 \cdot x = x \cdot b_1^a, \\ \vdots \\ b_k \cdot x = x \cdot b_k^a. \end{cases}$$

Then, set a basis for the solution space, and pick random solutions x where x is invertible. Then solve the following system of equations with the constraint $y \in C(C(b_1, \ldots, b_k))$.

$$\begin{cases} a_1 \cdot y = y \cdot a_1^b, \\ \vdots \\ a_k \cdot y = y \cdot a_k^b. \end{cases}$$

Finally, set a basis for the solution space, and pick random solutions y where y is invertible.

5.4. Linear decomposition attack

This section is devoted to survey the main ideas of linear decomposition attacks. Myasnikov and Roman'kov in [**161**] present this new attack on several known group-based cryptosystems. This attack gives a polynomial time deterministic algorithm that is able to find the shared secret key from the public data.

The linear decomposition attack is divided in two steps which are described below. The first step is to compute a basis of a vector space. The second step utilizes the basis found in the first step to determine v^{ab} from v, v^a, v^b.

Step 1: Finding a basis. Consider a finite-dimensional vector space V over the field F with basis $B = \{v_1, \ldots, v_r\}$ and the semigroup $\mathrm{End}(V)$ of endomorphisms of V. The elements of V are written as a linear combination of elements of B, and elements of $\mathrm{End}(V)$ are written as matrices relative to B. For any subset $W \subseteq V$ and $A \subseteq \mathrm{End}(V)$ we write $W^A = \{w^a \mid w \in W, a \in A\}$. Furthermore, we denote by $Sp(A)$ the subspace of V generated by W and by $\langle A \rangle$ the submonoid generated by A in $\mathrm{End}(V)$.

The following result yields an algorithm to find a basis for $Sp(W^{\langle U \rangle})$ in polynomial time.

LEMMA 5.1. [**161**, Lemma 2.3] *There exists an algorithm that for given finite subsets $W \subseteq V$ and $U \subseteq \mathrm{End}(V)$ finds a basis of the subspace $Sp(W^{\langle U \rangle})$ in the form $w_1^{a_1}, \ldots, w_t^{a_t}$, where $w_i \in W$ and a_i is a product of elements of U.*

The main idea of the algorithm, which is provided in the proof (see [**161**]), is to use Gaussian elimination to determine a maximal linearly independent set of W. Beginning with a linearly independent subset L_0 of W, one can successively add elements v^a with $v \in L_0$ and $a \in U$ and check for linear independence with each addition. This will create a chain of strictly increasing subspaces of V, and since V is finite dimensional this chain must eventually terminate.

Step 2: The basic linear decomposition attack. Let V, as before, be a finite-dimensional vector space over a field F with basis $B = \{v_1, \ldots, v_r\}$ and let U and W be two finite subsets of $\mathrm{End}(V)$. In **[161]** the authors assume furthermore that every element of U commutes with every element of W. Consider A and B the submonoids of $\mathrm{End}(V)$ generated by U and W, respectively. The data is organized as follows:

- Public: the field F, the space V, the sets U, W and the vectors v, v^a, v^b.
- Private: $a, b \in End(V)$.

Given U, W, v, v^a and v^b one can find in polynomial time $v^{ab} = v^{ba}$ without determining a or b. Using the lemma above, one can determine a basis, $\{v^{a_1}, \ldots, v^{a_n}\}$ for the subspace $Sp(v^A)$ given U and v. One can write v^a as a linear combination in the determined basis, which in turn allows one to compute v^{ab}. To see this note that if $v^a = \sum_{j=1}^{n} \alpha_i v^{a_i}$ with $\alpha_i \in F$, we have that

$$v^{ab} = (v^a)^b = (\sum_{j=1}^{n} \alpha_i v^{a_i})^b = \sum_{j=1}^{n} \alpha_i v^{a_i b} = \sum_{j=1}^{n} \alpha_i v^{ba_i} = \sum_{j=1}^{n} \alpha_i (v^b)^{a_i}.$$

Since v^b is public information, this value is computable. Hence, this provides a method to compute v^{ab} without knowing a or b. Notice that, in contrast with the classical case, we are not required in this case to solve the discrete logarithm-type problem of recovering a from v^a in order to compute the shared key.

Specific applications. In **[161]** several examples of applications of the linear decomposition attack are given. Let G be a finitely generated group equipped with an injective homomorphism $\psi : G \to GL_n(M)$, where M is a finite-dimensional associative algebra over a field F. Let U and W be two finite subsets of G that satisfy as before the law $[U, W] = 1$. Consider A and B submonoids of G generated by U and W, respectively.

- A linear group acting by conjugation:
  Suppose that $a \in A$, $b \in B$, and $v \in G$. Then, given U, W, g, g^a, and g^b, one can find in polynomial time the element $g^{ab} = g^{ba}$.
- A linear group acting by right/left multiplication:
  Suppose that $a, a' \in A$, $b, b' \in B$, and $v \in G$. Then, given U, W, g, agb, and $a'gb'$ one can find in polynomial time the element $a'agbb' = aa'gb'b$.

5.5. Quotient attack

In this section we discuss the quotient attack (denoted QA). This attack provides a way to solve some decision problems in groups more efficiently. As the name suggests, the idea of a quotient attack is to reduce the problem to a quotient G/N of a given group G. These algorithms were initially studied in **[125]** in the case of decision problems. It is important to note that the method cannot be applied for search problems. In **[164]** an approach with the additional hypothesis that the group has some version of the free basis property was studied. Below we give an idea of how QA works.

Let $G = \langle X \rangle$ be a finitely generated group and consider the canonical epimorphism $\psi : G \to G/N$. Let $H = \langle u_1, \ldots, u_k \rangle$ be a finitely generated subgroup of G.

The following algorithm gives a heuristic solution to the membership search problem (MSP). Let $F(X)$ be the free group over the alphabet X.

- Input: A word $w = w(X)$ and generators $\{u_1, \ldots, u_f\} \subseteq F(X)$ of a subgroup H are given.
- Output: A representation $W(u_1, \ldots, u_k)$ of w as an element of H or failure.

The algorithm works as follows:

- First, compute the generators $\psi(u_1), \ldots, \psi(u_k)$ of $\psi(H)$ in G/N.
- Compute $\psi(w)$ and solve the membership problem for $\psi(w)$ and $\psi(H)$. Find a representation $W(\psi(u_1), \ldots \psi(u_k))$ of $\psi(w)$ as a product of the generators of $\psi(u_1), \ldots, \psi(u_1)$ and their inverses.
- Check if $W(u_1, \ldots, u_k)$ is equal to w in G. If this is the case, the output is W; otherwise, the output is failure.

The algorithm above is deterministically correct if one is able to solve the word problem in G. More precisely, one first must solve the MSP in the quotient G/N and then check if the result works in G.

The authors in [**164**] showed that the CKEP (see section 5.3) can be reduced in linear time to the simultaneous conjugacy search problem relative to a subgroup. In addition, they proved the the simultaneous conjugacy search problem relative to a subgroup on an exponentially generic set of inputs is solvable in at most quadratic time relative to the time complexity of computing the geodesic length in the subgroup. Using this and the algorithm to the membership search problem above, one can solve the simultaneous conjugacy search problem relative to a subgroup on an exponentially generic set of inputs. We refer the reader to [**164**] for details.

CHAPTER 6

Quantum Cryptanalysis

6.1. Introduction

In this chapter, we will see the application of quantum methods to group-theoretic computational problems. Of course, certain instances of the *hidden subgroup problem* are famously quantum-vulnerable, as there exist polynomial-time quantum algorithms providing their solution. We will review the main ideas behind these algorithms, and the groups for which the hidden subgroup problem remains out of reach even for a quantum-equipped adversary. We then move onto a survey-type report of computational problems in cryptography that reduce to some instance of the hidden subgroup problem, before finishing with two "case study"-type examples detailing the relatively recent quantum analysis of two group-theoretic computational problems.

6.1.1. Quantum preliminaries. The purpose of this chapter is to treat with a reasonable level of detail the ideas used in theoretical quantum algorithms. As such, the reader should be familiar with some of the basic ideas of quantum computation, which we will review here.

A *qubit*, for our purposes, is a 2-dimensional vector space over the complex numbers $\mathbb{C}$, written[1] H_2. We will always consider the so-called *computational basis* of these systems; in particular

$$|0\rangle := \begin{bmatrix}1\\0\end{bmatrix} \qquad |1\rangle := \begin{bmatrix}0\\1\end{bmatrix}.$$

A general *state* ψ of a qubit can be represented by

$$\psi = \alpha\,|0\rangle + \beta\,|1\rangle\,,$$

where $\alpha, \beta \in \mathbb{C}$ are such that $\alpha^2 + \beta^2 = 1$. A crucial property of this system is that we can *measure* it; retrieving $|0\rangle$ with probability α^2 and $|1\rangle$ with probability β^2. The measurement process is in a sense *destructive*, in that one cannot deduce from a measurement outcome the state that was measured.

Two qubits taken together, meanwhile, can be represented by a 4-dimensional vector space over $\mathbb{C}$, written H_4. In this case the basis states are

$$|00\rangle = \begin{bmatrix}1\\0\\0\\0\end{bmatrix}, \quad |01\rangle = \begin{bmatrix}0\\1\\0\\0\end{bmatrix}, \quad |10\rangle = \begin{bmatrix}0\\0\\1\\0\end{bmatrix}, \quad |11\rangle = \begin{bmatrix}0\\0\\0\\1\end{bmatrix};$$

and a general state looks like $\alpha\,|00\rangle + \beta\,|01\rangle + \gamma\,|10\rangle + \delta\,|11\rangle$, where $|\alpha|^2 + |\beta|^2 + |\gamma|^2 + |\delta|^2 = 1$. In fact, this 4-dimensional state decomposes nicely with respect to

[1]The H here is for *Hilbert*; in fact, we are dealing with specifically a Hilbert space, but this detail is suppressed.

the *tensor product*, written $\otimes$, under which one has $|i\rangle \otimes |j\rangle = |ij\rangle$ (one might also see this written $|i,j\rangle$ or $|i\rangle\,|j\rangle$). One could more generally write that $H_4 = H_2 \otimes H_2$, where the basis states $\mathbf{z}$ of H_4 are exactly the vectors $|x\rangle \otimes |y\rangle$ such that $|x\rangle\,, |y\rangle$ are basis states of H_2. This idea generalizes as one might expect, and n qubits taken together can be represented by a 2^n-dimensional vector space. Such a system is called a *quantum register* of length n, and it is common to write its basis states with respect to their binary expansion; that is, the basis states are $\{|i\rangle : 0 \leq i < 2^n\}$.

It is also important to understand what happens when we measure these multi-qubit systems. Returning to the example of two qubits, consider the state $|\psi\rangle = \alpha\,|00\rangle + \beta\,|01\rangle + \gamma\,|10\rangle + \delta\,|11\rangle$. Measuring the whole system gives (with respect to binary representation) $0, 1, 2$ or 3 with respective probabilities $|\alpha|^2, |\beta|^2, |\gamma|^2, |\delta|^2$. We may, however, choose to measure only one of the qubits: say we measure the first one and record a $|0\rangle$. An important idea is that this collapses the system; since a 0 was observed, the system must be left in that state $1/N(\alpha\,|00\rangle + \beta\,|01\rangle)$, where N is the appropriate normalizing factor. In other words, the state of a system after partial measurement must be consistent with the outcome of that measurement.

6.2. The hidden subgroup problem

The hidden subgroup problem (HSP) asks the following: given a description of a group G and a function $f : G \to X$ that, for some finite set X, is constant and distinct on left (or right) cosets of a subgroup $H \leq G$, find a generating set for H. The group H is called the *hidden subgroup* for this problem.

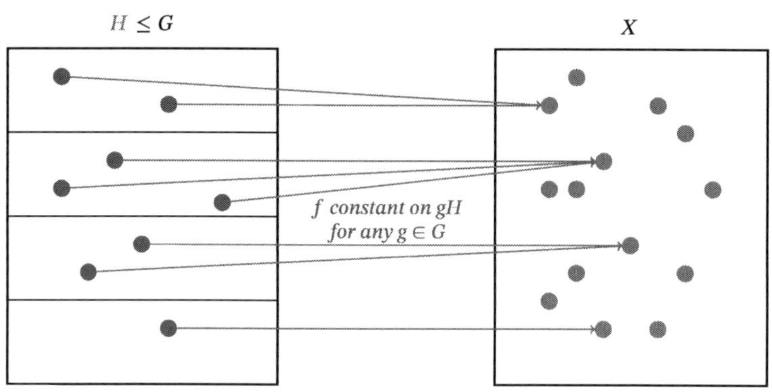

It is well known that computational hardness problems such as factoring and finding the discrete logarithm [**16**] are special cases of the hidden subgroup problem (HSP), and indeed that Shor's famous factoring algorithm is a special case of a more general algorithm addressing the HSP. Other cases in this vein can be found in the literature; for example, Kitaev's algorithm for the abelian stabilizer problem [**130**] solve a special case of the HSP. Indeed, the HSP is directly related to the discrete logarithm problem, graph isomorphism problem, lattice-based problems, and the problem of factoring via RSA.

6.2.1. Shor's algorithm. No discussion of quantum cryptanalysis can be complete without reference to Shor's algorithm. One of the consequences of Peter Shor's celebrated 1995 paper [**194**] is the very first instance of quantum cryptanalysis — indeed, since factoring and the discrete logarithm problem can very naturally be thought of as preliminary examples of group-based cryptography, we can think of Shor's work in modern terms as the first example of group-based quantum cryptanalysis.

In this short section we seek to provide some greater depth to the general notion that Shor's algorithm allows fast factoring, or that Shor's algorithm solves the hidden subgroup problem. In particular, we seek to give a short overview of what the quantum part of Shor's algorithm is "good" at, and the three inherently nonclassical ingredients that provide an advantage over classical approaches. The exposition is synthesized from a somewhat arbitrary subset of many excellent resources on the subject: in particular [**106**], [**169**], [**32**], [**173**].

It turns out the most illustrative example for our purposes is not directly the discrete logarithm problem, or factoring, but *order-finding* or *period-finding*; that is, the following task: given an integer N and an arbitrary integer a such that $1 < a < N$, find the smallest r such that $a^r \equiv 1 \mod N$. In other words, we would like to find the period of the function f defined by $f(x) = a^x \mod N$.

The first quantum ingredient is the ability to calculate all values of f *at once*, in the following sense. Any circuit[2] of classical gates computing a classical function can, by adding some redundancy, be converted into a quantum circuit. In particular, suppose $f : \{0,1\}^n \to \{0,1\}^m$ can be computed classically and efficiently for $n, m \in \mathbb{N}$; then there exists a quantum circuit Q_f such that

$$|x\rangle \underbrace{|0...0\rangle}_{m \text{ times}} \stackrel{Q_f}{\curvearrowright\!\!\!\longrightarrow} |x\rangle\,|f(x)\rangle\,,$$

where several 0-ancilla registers are suppressed. Suppose that we feed each of the m qubits comprising the first register first through a Hadamard gate (a gate that transforms a basis state into the uniform superposition of all basis states). One can check that the result is:

$$|x\rangle\,|0\rangle \quad \underset{\text{Hadamard gates}}{\overset{\text{1st register through}}{\curvearrowright\!\!\!\longrightarrow}} \quad \frac{1}{\sqrt{2^n}}\sum_{i=0}^{2^n-1} |x\rangle\,|0\rangle \quad \stackrel{Q_f}{\curvearrowright\!\!\!\longrightarrow} \quad \frac{1}{\sqrt{2^n}}\sum_{i=0}^{2^n-1} |x\rangle\,|f(x)\rangle\,.$$

In other words, with a single application of a quantum circuit no less efficient than an equivalent classical circuit, one can obtain the superposition

$$(*) \qquad \frac{1}{\sqrt{N}}\sum_{i=0}^{N-1} |i\rangle\,\big|a^i \mod N\big\rangle\,.$$

The next ingredients are the laws of partial measurement, and the ability of a quantum computer to perform a Fourier transform quickly. Drawing now from [**35**, Algorithm 2] — which presents the order-finding routine slightly differently to its original form — if we measure[3] the second register, the first register must be left in a state consistent with the outcome of this measurement. More concretely,

[2]Where *circuit* loosely means a configuration of finitely many gates

[3]The principle of implicit measurement [**169**, p.187] actually gives that we do not need to perform this measurement at all — we can simply discard the second register. Nevertheless, understanding what happens when we do this is best understood by imagining that we have indeed measured the appropriate register.

if measuring the second register of the state $(*)$ yields a value $a^l \mod N$, without loss of generality the first register is left in the state

$$(**) \qquad \frac{1}{\sqrt{s}} \sum_{i=0}^{s-1} |l + ir\rangle \,,$$

where r is the smallest integer such that $a^r \equiv 1 \mod N$ and s is some normalizing factor.

Intuitively we have somehow *trapped* some information about the order r here, but we are not yet done — note that we do not a priori know the values of s or l, so measurement at this stage does not help. Here, the quantum Fourier transform comes into play. This fundamental aspect of quantum computation will be treated here as a black box that somehow amplifies the probability of observing a *useful* value; in particular, when applied to $(**)$, yielding

$$\frac{1}{\sqrt{r}} \sum_{k=0}^{r-1} \omega_r^{lk} \left| \frac{kN}{r} \right\rangle,$$

where ω_r is the rth root of unity. With reasonable probability, measuring this state gives a value which allows successful recovery of r after some classical post-processing.

The order-finding ability we gained with distinctly quantum tools is, in a sense, the crux of quantum algorithms that solve many instances of the hidden subgroup problem (HSP). Factoring, for example, is well known to reduce to a problem of order-finding; the algorithm given for solving the discrete logarithm problem in [**194**] is a slightly more involved application of very similar ideas. More generally, one can extend these ideas to efficiently solve the HSP for any finite abelian group.

In what remains of this section let us survey some of the results for groups falling outside of this class.

6.2.2. Nonabelian finite groups. If the group G is either solvable or a permutation group, and the hidden subgroup H is normal, an algorithm is presented in [**114**] to find hidden subgroups of groups with small commutator subgroup and of groups admitting an elementary abelian normal 2-subgroup of small index or with cyclic factor group. Moreover, recursive algorithms to solve the HSP in some classes of nonabelian solvable groups have been proposed in [**66**].

Now we survey what has been proved in some finite nonabelian groups to solve the HSP. Namely, we address here the cases of the Heisenberg group, the dihedral groups and a specific wreath product (see Chapter 1 for more information on these groups).

Let G be the Heisenberg group, and let $H_p = \langle (a,b,c) | a,b,c \in \mathbb{Z}_p \rangle$, where

$$(a_1, b_1, c_1)(a_2, b_2, c_2) = (a_1 + a_2, b_1 + b_2, c_1 + c_2 + a_1 b_2).$$

Let $H_{a,b} = \langle (a,b,1) \rangle = \{(a,b,1)^x | x \in \mathbb{Z}_p\}$. To find an arbitrary subgroup of G reduces to determining two parameters a, b, given the coset state produced by the standard method with the f which hides $H_{a,b}$. To this end, it is enough to distinguish cyclic subgroups of order p of the above form. This can actually be computed with a success probability close to $\frac{1}{2}$.

The dihedral group D_n of order $2n$ is given by the presentation $D_n = \langle x, y | x^n = y^2 = yxyx = 1 \rangle$. In this case the situation is a bit more complicated and the function f hides subgroups H of order 2. Kuperberg ([**139**]) proved that to find an arbitrary hidden subgroup H of D_n reduces, in some sense, to finding the slope of a reflection of 2-dimensional n-gon. The algorithm provided to achieve this task has subexponential time and query complexity. Regev's algorithm [**181**] runs with polynomial space but slightly worse complexity.

The case in which G is a wreath product of the form $G = \mathbb{Z}_2^n \wr \mathbb{Z}_2$ was studied in [**187**]. Here, the authors provide a positive result for finding an efficient solution to the nonabelian HSP.

Observe that in most cases, the inefficiency of a proposed quantum algorithm strongly depends on the inefficiency of the required quantum measurement or post-processing within the group.

6.2.3. Infinite groups. We finish by considering the case of infinite groups. Recall that the process of the state is given by

$$|G\rangle = |G|^{-1/2} \sum_{g \in G} |g\rangle.$$

In the case of infinite groups, the first obstacle is that the quantum computer should assume the state above, whose meaning is not clear in the case in which the order of the group G is not finite. A possible way to work with infinite groups is to find a good way to encode the elements of the group. For example, elements of $\mathbb{Q}$ may be encoded as fractions, and elements of a free group as reduced words.

The abelian infinite hidden subgroup problem was first considered with Shor's algorithm over the integers $\mathbb{Z}$ with addition, and in [**77**], this problem has been explored for infinite abelian groups $\mathbb{Z}^N$. Later on, the HSP was studied for groups of the form $\mathbb{R}^k \times \mathbb{Z}^l \times (\mathbb{R}/\mathbb{Z})^s \times H$, where H is a finite group, [**48**].

In a very recent preprint, Kuperberg showed that if $G = (\mathbb{Q}, +)$ or $G = F_k$, where F_k denotes the free group of rank k, then the hidden subgroup problem is NP-hard.

Other than these cases mentioned above, very little is known in the literature about the HSP problem for infinite groups.

6.3. Relation of HSP to other computational problems

In this section we examine the relationship between HSP and other important computational problems. Indeed, as pointed out in the introduction of this chapter, an efficient algorithm for HSP over a certain group produces an efficient algorithm for other computational problems such as the discrete logarithm problem or the graph automorphism problem.

First of all, one has to determine an appropriate group G, the subgroup H and the function f. Note that we require the function f to be H-periodic, which means that it is constant and distinct on left (or, respectively, right) cosets of a subgroup $H \leq G$. In 1994 Simon provided one of the first examples for which a quantum computer is much faster than a classical one. Simon's problem can be seen as a HSP over the group $\mathbb{Z}_2^n$ with a subgroup H of order 2. Shor's algorithm solves a special case of HSP, giving rise to an efficient quantum algorithm for factoring and for the discrete logarithm problem.

Another example of a problem that gives an instance of the HSP is the graph automorphism problem. Let Γ be a graph with n labeled vertices $\{v_1, \dots v_n\}$, S_n the symmetric group over n letters. Let X be the set whose elements are the collections $\{v_{\sigma(1)}, \dots, v_{\sigma(n)}\}$, for every permutation $\sigma \in S_n$. Now consider the function $f : S_n \to X$ that takes σ to $\{v_{\sigma(1)}, \dots, v_{\sigma_n}\}$, and call Γ_σ the graph Γ after relabeling its vertices according to σ. Then, the hidden subgroup H for (S_n, X, f) is exactly the group of automorphisms of Γ, and then the HSP for this triple is equivalent to finding a generating set of $Aut(\Gamma)$, which is a formulation of the graph automorphism problem. In a similar fashion, the graph isomorphism problem is an instance of HSP over the wreath product $G = S_n \wr S_2$. See [**33**] for more details, and [**134**] for a general introduction of these graph problems.

In turn, it has been shown in [**16**] that a solution to a particular instance of HSP is a solution to the problem of hidden linear functions. This problem says that if g is a permutation of $\mathbb{Z}_N$ and we have the function

$$\begin{aligned} h : \mathbb{Z} \times \mathbb{Z} &\to \mathbb{Z}_N \\ (x, y) &\mapsto x + ay \mod N, \end{aligned}$$

the map $f = g \circ h$ hides the subgroup $\langle(-a, 1)\rangle$.

Another interesting application was shown by Regev [**181**], who connected the solution to the dihedral hidden subgroup problem to a quantum solution to lattice problems. Specifically, the $g(n)$-unique shortest vector problem (USVP) is NP-hard for $g(n) = \mathrm{O}(1)$, and has a polynomial time classical solution when $g(n)$ is large. A solution to the dihedral hidden subgroup problem based on the standard-method can be used to solve poly(n)-USVP. The hidden subgroup problem over the symmetric and dihedral groups are highly motivated open questions in post-quantum group-based cryptography.

Another related computational problem is the hidden shift problem, which has been proposed as a basis for post-quantum cryptography in symmetric cryptosystems that are quantum-CPA secure [**1**]. Other than the use of a generalization of Simon's algorithm, and Kuperberg's algorithm discussed above, very little is known about the hidden shift problem. As some solutions coincide, clearly this problem is closely related to the hidden shift problem. It is important to note that constructions based on the hidden shift problem have also remained quantum secure.

Additionally, self-shift-equivalent polynomials can be framed as an instance of HSP; in this case, Grigoriev shows how to compute the hidden subgroup [**89**]: given a polynomial $P \in \mathbb{F}_q[x_1, \dots, x_l]$, we define f such that $(a_1, a_2, \dots, a_l) \mapsto P(X1 - a_1, X_2 - a_2 \dots, X_l - a_l)$ is constant on cosets of $H \leq \mathbb{F}_q^l$. This subgroup is the set of self-shift-equivalences of the polynomial P, and Grigoriev shows how to compute the hidden subgroup [**89**].

6.4. Quantum cryptanalysis for group-based PQC

Let us conclude this chapter with two "case study"-type analyses of proposed post-quantum hard group-theoretic computational problems for which concrete quantum algorithms are known. Both problems seek to generalize the usual notion of the discrete logarithm problem (DLP) in such a way as to not admit a reduction to an easy instance of the HSP.

6.4.1. The semigroup discrete logarithm problem. Let G be a cyclic group with generator $\langle g\rangle$. We consider the following facts to be folklore: DLP in this instance can be reduced to an easy HSP incidence, and that the construction of the hiding function relies on the existence of a group inverse. It is therefore a natural question to ask whether DLP is difficult in a semigroup. Without diving into the particulars of the rich subject of semigroup theory, we can safely assume in this work that *semigroup* means a finite group for which each group element is not guaranteed to have an inverse.

Let us formalize the question a little further.

DEFINITION 6.1. *Let G be a finite semigroup. For any $g \in G$, the set $\langle g \rangle := \{g^i : i \in \mathbb{N}\}$ is called the monogenic semigroup generated by g.*

In this language, we would like to know if, in a reasonable monogenic semigroup (say, one in which the semigroup multiplication is efficient) the following problem is easy: given g and g^x for some $x \in \mathbb{N}$, recover the integer x. Intuitively, we might expect this to be difficult since, as alluded to above, we can no longer compute the hiding function in the reduction to HSP. It turns out, however, that this problem admits a fast quantum algorithm due to the structure of monogenic semigroups.

LEMMA 6.2. *Let G be a finite semigroup. To each $g \in G$ is associated a pair of integers n, r such that*

$$\langle g \rangle = \{g, ..., g^n, ..., g^{n+r-1}\}$$

and $g^r g^i = g^i$ for each $i \geq n$. Moreover, the set $\{g^n, ..., g^{n+r-1}\}$ is a cyclic group under the semigroup operation.

Sketch of proof. This fact is rather standard and can be found in, say, [**111**]. The main idea of the proof is to note, since G is finite and $\langle g \rangle \subset G$, that $\langle g \rangle$ is itself finite. It follows from the pigeonhole principle that there is a smallest integer n for which some $m > n$ is such that $g^n = g^m$. Writing $m = n + r$ for some $r \in \mathbb{N}$, the fixing property of g^r follows immediately. Carefully checking the appropriate minimality conditions shows that $g^i \in \{g^n, ..., g^n + r - 1\}$ for each $i \geq n$; to show that this set is a group, it remains only to show the existence of inverses and an appropriate generator, which is a simple matter of computation in the exponents. □

The integers n and r are called the *index* and *period* of g. Supposing that we know these two values, deriving the value x from g^x is easy; first, we determine if $x \geq n$ by checking whether $g^r g^x = g^x$. If indeed this is the case, we are left with a simple DLP; otherwise, since the set of all powers of g is in a sense *well-ordered* according to their invariance under multiplication by g^r, we are done by binary search in time $\mathcal{O}(\log n)$.

This strategy is the central idea of [**34**]. It remains to demonstrate that one can recover the index and period efficiently — in the same work, a quantum algorithm for precisely this task is presented.

LEMMA 6.3 (Lemma 1, [**34**]). *Let G be a semigroup and $g \in G$. There is an efficient quantum algorithm to recover the index and period of g.*

Sketch of proof. The algorithm is very similar to Shor's method of order-finding, though a slight loss in success probability is incurred. Under the assumption that multiplication in G is efficient, we can calculate powers of any $g \in G$ efficiently with square-and-multiply techniques. In other words, the usual technique of preparing the state

$$\sum_{i=0}^{N-1} |i\rangle \left|g^i\right\rangle$$

can be implemented efficiently with l qubits, where $N = 2^l$ is sufficiently large.

Suppose we measure the second register. With probability $(N-n)/N$, we observe an element g^i with $i \geq n$; we have seen that the integers j such that $g^j = g^i$ are of the form $x_0 + kr$. Supposing that there are M values of k such that the value of $x_0 + kr \leq N$, we are left with the state

$$\frac{1}{\sqrt{M}} \sum_{k=0}^{M} |x_0 + kr\rangle ,$$

upon which the same techniques appearing in Shor's algorithm can be applied, yielding the period with constant probability. □

To summarize, we have an example of a computational problem that does not appear to be significantly harder than HSP for finite abelian groups. Let us now see an example of a computational hardness problem whereby the gap between its difficulty and the difficulty of the finite abelian HSP appears to be nontrivial.

6.4.2. The semidirect discrete logarithm problem. Recall the HKKS key exchange protocol presented in Section 4.1.6.1. In this protocol we are interested in semigroup products of the form $\phi^{x-1}(g)...\phi(g)g$, where g is a semigroup element and ϕ is an endomorphism of the semigroup[4]; indeed, the security of the protocol is heavily related to the following.

DEFINITION 6.4. *Let G be a finite nonabelian semigroup and $End(G)$ be its endomorphism group. Given a pair $(g, \phi) \in G \ltimes End(G)$ and an element of G of the form*

$$\phi^{x-1}(g)...\phi(g)g$$

for some $x \in \mathbb{N}$, find the integer x.

At first flush it is not clear what relationship this problem has to the HSP. Nevertheless, in [**11**] it is shown that the problem admits a reduction to HSP for dihedral groups. We give the thrust of this argument below.

LEMMA 6.5. *Let G be a finite nonabelian semigroup and let $(g, \phi) \in G \ltimes End(G)$. We define the following quantities:*

$$s_{g,\phi}(x) = \phi^{x-1}(g)...\phi(g)g \quad \mathcal{X}_{g,\phi} = \{s_{g,\phi}(i) : i \in \mathbb{N}\}.$$

The following facts hold:

(a) *There exists a function $*$ such that for any $j \in \mathbb{N}$ and $s_{g,\phi}(i) \in \mathcal{X}_{g,\phi}$*

$$j * s_{g,\phi}(i) = s_{g,\phi}(i+j).$$

(b) *There exists a pair of integers (n, r) such that*

$$\mathcal{X}_{g,\phi} = \{g, ..., s_{g,\phi}(n), ..., s_{g,\phi}(n+r-1)$$

and $r * s_{g,\phi}(i) = s_{g,\phi}(i)$ for each $i \geq n$.

(c) *The cyclic group $\mathbb{Z}_n$ acts freely and transitively on the set $\{s_{g,\phi}(n), ..., s_{g,\phi}(n+r-1)\}$.*

[4] As presented, the protocol admits the use of a group and its automorphism group but allows relaxation to cases where invertibility is not available. In fact, the mechanics of what to follow look rather different depending on if we do or do not allow for invertibility. In this work we will review the noninvertible case since a majority of the proposed platforms suggested for use with the HKKS protocol do not allow invertibility.

Sketch of proof. The details of these facts are dealt with at some length in [**11**]. The existence of the $*$ operator is since

$$\phi^j(s_{g,\phi}(i))s_{g,\phi}(j) = \phi^{i+j-1}(g)...\phi(g)g = s_{g,\phi}(i+j).$$

In other words, the $*$ operator is just another way of phrasing the method of key agreement in the HKKS protocol. The existence of the pair of integers n, r is similar to the semigroup case discussed above; however, unlike in the semigroup case, we cannot directly multiply elements of $\mathcal{X}_{g,\phi}$ together and obtain an element in $\mathcal{X}_{g,\phi}$ because the semigroup is noncommutative. The definitions of n, r, and $*$ immediately give rise to the invariant property of r; a matter of careful checking rather intuitive notions allows one to verify the group action. □

Recall that in the semigroup context, modulo a tail of values before the looping behavior started, we were left more or less with a cyclic group. In this case the structure is described instead by a group *action*, and the relationship to the HSP is less clear. Indeed, suppose again that we know the values n, r corresponding to a pair (g, ϕ), and we are given $s_{g,\phi}(x)$. In order to recover the integer x we can first verify whether $x \geq n$ by checking if $r * s_{g,\phi}(x)$; if not, we are done by binary search and repeated application of the $*$ operator. Otherwise, without loss of generality[5] we are left with the following task: recover the integer y such that $y * s_{g,\phi}(n) = s_{g,\phi}(x)$.

Fortunately, this task is rather well-studied and is defined (for group actions) by Couveignes as the *Vectorization Problem* [**36**]. For our purposes, we will be particularly interested in the easy reduction of this problem to the abelian hidden shift problem — it is well known, say in [**139**], that this problem is exactly the dihedral hidden subgroup problem.

Let us see that our problem relating to group actions is indeed the abelian hidden shift problem.

DEFINITION 6.6. *Let A be an abelian group and S a set. Two injective functions $f, g : A \to S$ hide a value $s \in S$ if for each $a \in A$ one has $f(a) = g(a + s)$. The abelian hidden shift problem asks one to recover the value s.*

Translated to our context, the abelian group in question is $\mathbb{Z}_r$ and the set is $\{s_{g,\phi}(n), ..., s_{g,\phi}(n + r - 1)\}$ (written for brevity as $\mathcal{C}_{g,\phi}$, where for obvious reasons the $\mathcal{C}_{g,\phi}$ stands for *cycle*). Defining $f, g : \mathbb{Z}_r \to \mathcal{C}_{g,\phi}$ by

$$f([a]_r) = [a]_r * s_{g,\phi}(x) \quad g([a]_r) = [x]_r * s_{g,\phi}(n),$$

where the group action $*$ is the natural extension to residue classes, it is easy to see that these functions hide the value y such that $n + y = x$, since

$$\begin{aligned} f([a]_r) &= [a]_r * s_{g,\phi}(n+y) \\ &= [a]_r * ([y]_r * s_{g,\phi}(n)) \\ &= [a+y]_r * s_{g,\phi}(n) \\ &= g([a+y]_r). \end{aligned}$$

Our understanding of the relationship of the semidirect discrete log problem to the HSP is therefore complete provided the knowledge of n and r. To finish, let us see that an efficient quantum algorithm recovers these values.

[5]That is, we are technically looking for a residue class, but its smallest positive representative will do.

LEMMA 6.7 (Theorem 4,[**11**]). *There exists an efficient quantum algorithm for recovering the values* (n, r) *corresponding to a pair* $(g, \phi) \in G \ltimes End(G)$.

Sketch of proof. The quantum part of the algorithm is very similar to the one described in Lemma 6.3; we can think of replacing the usual exponentiation function with the $s_{g,\phi}$ function. Since this function also displays r-periodic behavior for integers larger than n, partial measurement of an analogous state gives a superposition on which one can use the standard tools to recover r. The integer n is then recovered clasically, again with binary search since the integers are well-ordered with respect to invariance under $r*$. □

In this latter case, then, we appear to have a nontrivial gap in quantum hardness with respect to the standard DLP, since the time complexity of Kuperberg's algorithm is subexponential. Similar ideas employing group *actions* rather than groups are common in isogeny-based cryptography, and appear, for example, in [**26**].

Part 3

Cryptographic Platforms

CHAPTER 7

Braid Groups

7.1. Introduction

In this portion of the text, we start the analysis, from the algorithmic and cryptographic point of view, of different distinguished families of groups. In every chapter, we will introduce the family using a combinatorial approach in order to make the subject accessible for non-group-theoretic specialists, and also because the definitions in terms of generators and relations are usually the most convenient from the point of view of cryptographic protocols. We intend to make every chapter self-contained from the mathematical point of view, avoiding unnecessary extra information and providing only the definitions and results that are essential to understand the decision problems and the cryptographic platforms.

We start by analyzing the family of *braid groups*. These groups are probably the most important examples of the family of Artin-Tits groups, and were discovered by Emil Artin himself in 1925. They admit different definitions; for example, as fundamental groups of configuration spaces, mapping class groups of surfaces, or in terms of the composition of geometric braids. Because of this, they can be considered as a crossroad of different branches of mathematics, and possess a rich structure. In particular, they have been thoroughly studied from the algorithmic point of view and they have been considered as a platform for different cryptographic protocols. However, their role in this context is currently controversial, as they have received some criticism because of security issues.

We briefly describe the structure of this chapter. We start with the main definitions concerning braid groups, and some invariants and previous results that are required later. Then, we will review different solutions that have been proposed for the main decision problems for braid groups. In the last section we will describe how these algorithmic properties are applied in cryptographic applications and analyze their security.

As mentioned above, we present here only the necessary material to follow our exposition. There are excellent surveys in braid theory, such as [**79**] or [**126**], where, in particular, the reader may find the proofs of the results quoted in this section.

7.2. Mathematical background

We begin with the classical definition of braid groups in terms of Artin generators and relations.

DEFINITION 7.1. *Let $n \geq 2$. The* braid group B_n *is defined by the following presentation:*

$$\{\sigma_1, \dots, \sigma_{n-1} \mid \sigma_i\sigma_j\sigma_i = \sigma_j\sigma_i\sigma_j \textit{ if } |i-j| = 1, [\sigma_i, \sigma_j] = 1 \textit{ if } |i-j| > 1\}.$$

This is called the *Artin presentation*, and the σ_i are called the *Artin generators*. The elements of the group are called *braids*, as they can be represented by geometrical braids, and the product by concatenation of them.

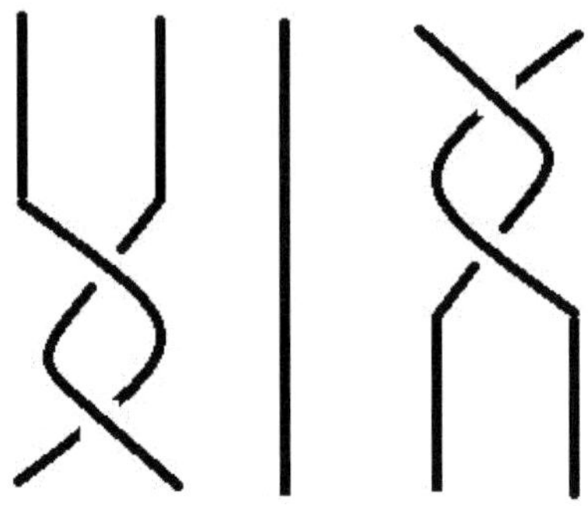

FIGURE 7.1. The braid $\sigma_4^2\sigma_1^2$ in B_5.

Although in general we will use only the combinatorial definition of braids in this chapter, the notion of geometric braid is more appropriate to define the key notion of *pure braid.* We explain this concept, starting with the definition of *strand.*

DEFINITION 7.2. *A* strand *is a path* $\gamma : [0,1] \to \mathbb{C} \times [0,1]$ *that increases monotonically in the second coordinate.*

Now we fix a set of n base points in $\mathbb{C}$, say the first n integer numbers. Consider a set of n disjoint strands such that:

- The i-th strand starts at $(i,0)$ and finishes at $(j,1)$ for some $j \in \{1,\dots,n\}$.
- Two different strands finish in different points.

Between any two such sets β and β' of n strands, we consider the following equivalence relation: $\beta \sim \beta'$ is and only if they are isotopic, and the isotopy fixes $\{1,\dots,n\} \times \{0\}$ and $\{1,\dots,n\} \times \{1\}$ pointwise. Under these conditions, we have the following definition:

DEFINITION 7.3. *A* geometric n-braid *is an equivalence class by the previous relation.*

Observe that the isotopy condition means that that two collections of strands represent the same braid if we can deform continuously one into the other inside $\mathbb{C} \times [0,1]$ without moving the endpoints.

Given two geometric braids, we can define a product by concatenation (after stacking and rescaling). With this operation, the set of n-geometric braids forms a group, which is isomorphic to B_n.

The braid group can also be defined as the fundamental group of a configuration space, and also as a mapping class group. We do not deal with these definitions here; the interested reader is referred to [**79**].

Observe that every braid β determines a permutation τ such that $\tau(i) = j$ if the i-th strand of β finishes at the point $(j,1)$. This defines a homomorphism $\pi : B_n \to S_n$, and we have:

DEFINITION 7.4. *The kernel of* π *is usually denoted* PB_n, *and its elements are called* pure braids.

Note that pure braids are braids such that for every $i \leq n$, its i-th strand finishes at the point $(i, 1)$. The group PB_n is fundamental in the theory of braid groups, and in particular on the algorithmic side as we will see in the next section.

In [**14**] Birman gave a presentation of the pure braid group, which is stated in the proposition below.

PROPOSITION 7.5. *The pure braid group PB_n has a presentation whose generators are A_{ij}, with $1 \leq i \leq j \leq n$, and the following relations:*

- $[A_{rs}, A_{ij}] = 1$, *if* $i < j < r < s$ *or* $r + 1 < i < j < s$.
- $A_{ij}^{-1} A_{js} A_{ij} = A_{is} A_{js} A_{is}^{-1}$, *if* $i < j < s$.
- $Aij^{-1} A_{is} A_{ij} = A_{is} A_{js} A_{is} A_{js}^{-1} A_{is}^{-1}$, *if* $i < j < s$.
- $Aij^{-1} A_{rs} A_{ij} = A_{is} A_{js} A_{is}^{-1} A_{js}^{-1} A_{rs} A_{js} A_{is} A_{js}^{-1} A_{is}^{-1}$, *if* $i + 1 < r < j < s$.

PROOF. To see that this presentation defines the pure braid group, let

$$A_{ij} = \sigma_{j-1}\sigma_{j-2} \dots \sigma_{i+1}\sigma_i^2\sigma_{i+1}^{-1} \dots \sigma_{j-2}^{-1}\sigma_{j-1},$$

where the σ_i's are Artin generators. See [**14**]. □

DEFINITION 7.6. *Given a braid s, its* length $|s|$ *is the minimum number of letters needed to represent s in terms of the Artin generators.*

The following result is implicit in many applications of braid theory:

PROPOSITION 7.7 ([**79**], Section 2). The braid groups are torsion-free.

In particular, this is a necessary condition for a group to be *left-orderable*. A group G is said to be left-orderable if there is a total ordering $<$ in the elements of the group such that for every triple $x, y, z \in G$, $y < z$ implies $xy < xz$ (i.e., the order is *left-invariant*). We have:

PROPOSITION 7.8 ([**79**], Theorem 2.8). *The braid groups are left-orderable.*

Sketch of proof. A total order in the braid group B_n was defined by Dehornoy in the following way. We call a word w in the Artin generators σ_i-positive if:

- For every letter σ_j that appears in the expression of w, $j \geq i$.
- The letter σ_i appears in w, and its inverse does not.

Define $\alpha < \beta$ if there exists a representative for $\alpha^{-1}\beta$ that is σ_i-positive for some i. This order is left-invariant by definition, and it is proved in [**41**] that it is a well-defined total order in B_n. Thus, B_n is left orderable.

The order in the braid groups is crucial in many applications, and in particular in some solutions of decision problems.

There is a related notion of positiveness, defined below, which will be important in the study of braid groups.

DEFINITION 7.9. *A braid β is called* positive *if it can be represented as a word in the Artin generators with only positive exponents.*

The positive braids form a submonoid of B_n, usually denoted B_n^+. Garside [**70**] proved that the Artin presentation is in fact a presentation of B_n^+ as a monoid and that every braid α can be written as the product of a positive braid and the inverse of a positive braid.

We finish this introductory section by describing some normal forms that have been found in the context of braid groups. Recall that by *normal form* it is understood, for every element g in a group G and a presentation $\langle S \mid R\rangle$ of G, there exists a unique reduced word in the generators S and their inverses that represents g.

We first describe Garside normal form. Consider $\alpha, \beta \in B_n^+$; it is said that α is *(left) divisible* by β if there exists another positive braid γ such that $\alpha = \beta\gamma$. Garside proved in the article [**70**] that with this notion of divisibility, two positive braids always possess a least common (right) multiple (lcm) and a greatest common (left) divisor (gcd), so B_n^+ has a lattice structure.

The (left and right) lcm of the Artin generators, in particular, is a distinguished element of B_n^+ that is denoted Δ_n and can be defined inductively by the following conditions:

- $\Delta_1 = 1$.
- $\Delta_{i+1} = \Delta_i\sigma_i\sigma_{i-1}\ldots\sigma_2\sigma_1$.

In particular, $\Delta_2 = \sigma_1$, $\Delta_3 = \sigma_1\sigma_2\sigma_1$, $\Delta_4 = \sigma_1\sigma_2\sigma_1\sigma_3\sigma_2\sigma_1$, etc. The following result highlights the importance of this element.

PROPOSITION 7.10 ([**79**], Theorem 4.2). *The cyclic subgroup generated by* Δ_n^2 *is the center of* B_n.

We are now ready to define the Garside normal form.

DEFINITION 7.11. *Let w be a word in the Artin generators of B_n (and their inverses). Then the* Garside normal form *of w is the decomposition $w = \Delta_n^p b$, being $p \in \mathbb{Z}$, b a positive word such that a) Δ_n is not a left divisor of b, and b) the word b is lexicographically minimal for the positive words for which condition a) holds.*

This is a normal form because, under these conditions, p and b are unique. See [**70**].

We will give a sketch of how to construct the Garside normal form. Let w be a word in B_n written in terms of the Artin generators and their inverses. As σ_i is a divisor of Δ_n for every i, there exist $\{x_1, \ldots, x_n\}$ in B_n^+ such that $\Delta_n = x_i\sigma_i$, and then $\sigma_i^{-1} = \Delta_n^{-1}x_i$. Now, in every occurrence of an inverse of an Artin generator σ_i in w, we substitute this inverse by $\Delta_n^{-1}x_i$. Taking into account that for every i $\sigma_i\Delta_n = \Delta_n\sigma_{n-i}$, we obtain that $\sigma_{n-i}\Delta_n^{-1} = \Delta_n^{-1}\sigma_i$. Applying this property iteratively every time that a sequence $\sigma_i\Delta_n^{-1}$ appears in w, we can "move to the left" all the occurrences of Δ_n, obtaining in this way a decomposition $w = \Delta_n^r a$, with a positive. Finally, it could happen that Δ_n divides a. In this case, we move this Δ_n to the left, obtaining $w = \Delta_n^{r+1}a'$. If necessary, we iterate this process, taking into account that this can be done only a finite number of times, because in every step the length of the positive braid in the right diminishes. After this iteration, we obtain a positive word b' for which condition a) in the previous definition holds. Now we finish by choosing the lexicographic minimal word b that represents the braid defined by b'.

A drawback of Garside's method is that to check whether a divides Δ_n or not, the computation of all the positive representatives of a is required; this makes the algorithm highly inefficient.

Next we will describe a refinement of the Garside normal form, the *left normal form*, which was described by El Rifai-Morton in [**49**]. We begin with the definition of simple braid.

DEFINITION 7.12. *A positive braid* β *is called* simple *if there exists another braid* $\alpha \in B_n^+$ *such that* $\Delta_n = \beta\alpha$.

Given $b \in B_n^+$, we can consider the left gcd b_1 of b and Δ_n. In particular, when $b = b_1 b'$ the positive braid b_1 is called the *maximal simple left divisor* of b. In this way, we can give the following definition.

DEFINITION 7.13. *Let* $(k, b_1, \dots, b_r)$ *be a sequence with* $k \in \mathbb{Z}$, $\{b_1, \dots, b_r\}$ *nontrivial simple braids in* B_n^+ *that are different from* Δ_n. *If for every* i, b_i *is the maximal simple left divisor of* $b_i b_{i+1} \dots b_r$; *then* $(k, b_1, \dots, b_r)$ *is called a* normal *sequence.*

We are ready now to define the left normal form.

PROPOSITION 7.14 ([**49**], Theorem 2.9). *The elements of* B_n *are in one-to-one correspondence with the normal sequences; in particular, the* normal form *for a braid* b *is given by* $b = \Delta^k b_1 \dots b_r$ *with* $(k, b_1, \dots, b_r)$ *being a normal sequence.*

It is important to observe that k can be negative. Moreover, the number r is usually called the *complexity* of the braid.

The definition depends on choosing in a systematic way a representative of every simple braid. This can be done because there is a one-to-one correspondence between simple braids in B_n and elements of the symmetric group S_n.

We also recall here the definitions of cycling and decycling, as they will appear in later in the text.

DEFINITION 7.15. *Let* $b \in B_n$ *be a braid and* $(k, b_1, \dots, b_r)$ *its left normal form. Let* $\phi : B_n \to B_n$ *be the automorphism that sends* σ_i *to* σ_{n-i}, *for every* $i \leq n$. *Then it is said that the braid*

$$\partial_+(b) = \Delta_n^k b_2 \dots b_r \phi_n^k(b_1)$$

has been obtained by cycling *from* b, *and the braid*

$$\partial_-(b) = \Delta_n^k \phi_n^k(b_r) b_1 \dots b_{r-1}$$

has been obtained by decycling *from* b.

Observe that both $\partial_+(b)$ *and* $\partial_-(b)$ *are conjugate to* b.

There are other normal forms, such as the normal form of Birman-Ko-Lee [**15**] which is based on the Birman generators, or the alternating normal form of Dehornoy [**40**], which is based on a previous version of Burckel [**23**]. The interested reader may follow the provided references.

7.3. Algorithmic problems

In this section we will describe different solutions that have been given to the word problem and the conjugacy problem, as well as some variants of them. The latter is especially important in cryptographic applications.

7.3.1. The word problem. In this section we briefly discuss different solutions of the word problem for braid groups.

Solution 1. We begin with the original Artin solution to the problem. Observe that since a disk $\mathcal{D}_n$ with n holes retracts with deformation to a wedge of n-circles, the fundamental group of $\mathcal{D}_n$ is isomorphic to the free group F_n. The braid group B_n acts via automorphisms in $\mathcal{D}_n$, so there is a group homomorphism

$B_n \hookrightarrow Aut(F_n)$, which in fact is injective. Now, suppose x is an element of B_n, ϕ_x its associated automorphism of F_n and $\{x_i\}_{i\leq n}$ is a basis of F_n. To check that x is trivial is equivalent to checking that $\phi_x(x_i) = x_i$, for $i = 1, \ldots, n$. This is solvable because the word problem is solvable in F_n. This solution has exponential complexity, and hence it is not efficient.

Solution 2. The next solution, also given by Artin, is based on the Birman presentation described in the previous section. Given the braid group B_n, it can be shown that there exists an isomorphism

$$PB_n = F_{n-1} \rtimes (F_{n-2} \rtimes \ldots \rtimes F_1).$$

The process of describing a braid of PB_n as an element of the iterated semidirect product of free groups is referred to as *braid combing* because of its interpretation in terms of geometric braid as a successive "combing" of strings one at a time. In this way, every pure braid admits a normal form in terms of the Birman generators, and its construction can always be terminated in exponential time. Hence, to check if a braid β is trivial, the algorithm is as follows: first, we consider the image of β under the homomorphism $B_n \to S_n$. If the image is nontrivial, we are done. If not, then β is pure, and we can obtain in exponential time its combed normal form, which is easily checked to be trivial. The process has been improved in order to lower its complexity to polynomial time [**80**].

Solution 3. The next solution provided by Garside is based on the Garside normal form described earlier. Consider a braid x, whose Garside normal form is $x = \Delta^p A$, with A a positive braid such that Δ is not a prefix of A. Then x will be trivial if $p = 0$ and A is the trivial word, as this the only positive representative of the trivial braid. Hence, in this framework, the complexity of solving the word problem is the complexity of finding the Garside normal form of a braid. Unfortunately, the Garside algorithm for the word problem is not efficient. Later, El Rifai-Morton [**49**] modified this Garside normal form by defining the left normal form explained earlier; with this new expression, it was shown in [**51**] that the complexity of the algorithm reduces to $O(l^2 n\log n)$ with l being the length of the original braid in B_n.

Other solutions. There have been several other solutions that have been proposed for the word problem in braid groups. A classical way to address the issue is to represent faithfully the braid group inside some matrix group. If one gets such a representation, then to check that a word is trivial amounts to check that a certain product of matrices gives the identity, which can be done in polynomial time. However, depending on the ring of matrices, this polynomial complexity can be prohibitive from a practical point of view. For example, if F is a free module over $\mathbb{Z}[t, t^{-1}]$ of dimension $n-1$, the *Burau* representation is given by a homeomorphism $B_n \to GL(F)$ which is injective for $n \leq 3$ and not injective for $n \geq 5$ (the case $n = 4$ is open). In turn, the *Lawrence-Krammer representation* represents B_n inside the linear group over a $\mathbb{R}[t, t^{-1}]$-module.

From a different point of view, Dehornoy develops in [**39**] his method of *handle reduction*, which takes advantage of the existence of a linear ordering in B_n to perform a systematic reduction of any braid; this procedure leads naturally to another braid (its *reduced decomposition*) for which it is immediate to decide if it is trivial or not. The author claims that, at least for long braids, this is a very efficient method to compute a positive (or negative) representative.

Finally, we mention the *Dynnikov coordinates*, which are associated to an action of B_n over the lattice $\mathbb{Z}^{2n}$. As the coordinates determine the braid, this invariant also gives a solution to the word problem, and its time complexity is quadratic. For more information see [**41**].

7.3.2. The conjugacy problem. One of the main algorithmic open problems in braid groups, which as we will see in next section have important consequences from the cryptographic point of view, is to find efficient solutions for the conjugacy and conjugacy search problems. Presently, no general solution is known for these problems whose complexity is less than exponential. However, the different refinements of the concept of *summit set* suggest that the construction of a polynomial algorithm that solves these two problems may not be completely out of reach. In the following we will review the main advances in this topic.

In his cornerstone paper which was already mentioned in the previous section [**70**], Garside defined the summit set of an element of the braid group.

Consider an element $\beta \in B_n$, and let $w = \Delta^m A$ its Garside normal form. Now we conjugate w by all the simple braids in B_n (i.e., the divisors of Δ_n in B_n^+), and call the words obtained whose power is $\geq m$ $w_1, \ldots, w_i$. Then we apply the same process for each of the w_i, and recursively to the words of power $\geq m$ that are obtained. Iterating the process, a finite set of elements $W = \{w_i\}$ that are conjugate to w is obtained.

Definition 7.16. *The set W of elements of maximum exponent obtained by the previous procedure is called the* summit set *of $\beta \in B_n$, and is usually denoted by $S(x)$.*

Garside proved that two braids are conjugate if an only if their summit sets are the same, and provided the previous algorithm to compute the summit set, which immediately leads to a solution of the conjugacy problem in these groups. However, using his method, the computation of the summit set has exponential complexity; thus, it is an inefficient method.

After Garside's work, various authors have found different refinements or variations of these ideas, obtaining smaller subsets of the summit set that are algebraically computable, and such that their equality for two braids imply that they are conjugate. We can quote the following.

- The *super summit set* $SSS(x)$ of El Rifai-Morton [**49**] consists of the elements of $S(x)$ of minimal canonical length. The algorithm involves the cycling and decycling process described in previous section. Begin with a braid x, which enables $SSS(x)$ to be computed. Then, another braid y is conjugate to x if and only if an element of $SSS(y)$ belongs to $SSS(x)$. The procedure was improved by Franco-González-Meneses [**65**].
- The *ultra summit set* $U(x)$ defined by Gebhardt in [**72**] is the subset of $SSS(x)$ whose elements are the braids for which iterated cycling get eventually periodic. It improves a previous definition by Lee, which also permitted decycling before getting into the period.
- The *set of sliding circuits of* x, denoted by $SC(x)$ was defined in [**73**] by Gebhardt and González-Meneses. They use a procedure called *cyclic sliding*, which consists of conjugation by distinguished prefixes to develop their algorithm. As in the previous examples, given x and y, the possible outputs of the algorithm are those that x and y are conjugate, in which case

they also identify conjugating elements. If they are not, $SC(x)$ is obtained. In this way both the conjugacy and the conjugacy search problem are solved. It should be noted that $SC(x) \subseteq U(x) \subseteq SSS(x) \subseteq S(x)$.

The computation of all these subsets of Garside's summit set has been shown to be exponential in the number of generators of the corresponding braid group, so it is an important open problem to find an efficient solution for these problems. However, Gebhardt claims that the ultra summit set gives an efficient solution except in a very small set of elements, at least for $n < 400$. Efficiency has been proved for small values of n. Hence, there is still hope to find a polynomial solution.

7.4. Cryptographic protocols

In this section we will describe different protocols that have been proposed whose security depends on the nonexistence of an efficient solution for the conjugacy problem or a variant of it. We follow the approach of Dehornoy [**40**].

7.4.1. Anshel-Anshel-Goldfeld protocol. This key exchange protocol [**2**] has become classical, and it is based loosely on the difficulty of solving the conjugacy search problem (see Chapter 4). Then braid groups are (so far) a good platform to implement it.

Let us explain the protocol. Given a certain braid group B_n, there are public braids $p_1, \dots, p_l, q_1, \dots, q_m$. Alice may have a secret word in l generators, and Bob may have another secret word in m generators. We call these words respectively u and v.

Now Alice computes $u(p_1, \dots, p_l)$, conjugates all the $q_i's$ by this element, and sends the results $\bar{q}_1, \dots, \bar{q}_m$ to Bob. Bob performs the analog operation with v instead of u and the $p_j's$ instead of the $q_i's$, sending the results $\bar{p}_1, \dots, \bar{p}_m$ to Alice. Now Alice computes

$$u(p_1, \dots, p_l)u(\bar{p}_1, \dots, \bar{p}_l)^{-1}$$

and Bob computes

$$v(\bar{q}_1, \dots, \bar{q}_m)v(q_1, \dots, q_m)^{-1}.$$

These two elements, which are equal, are the common key.

The security of the protocol depends on the complexity of the subgroup-restricted simultaneous conjugacy search problem in the case of braid groups, which is bounded below by the complexity of the conjugacy search problem.

7.4.2. Diffie-Hellman. A Diffie-Hellman–like protocol was proposed by Ko et al. in [**131**], taking the advantage that every Artin generator in a braid group commutes with almost the rest of these generators. Then it is easy to find pairs of *big* subgroups of B_n such that any element of each of them centralizes the other. Now, if m is the integer part of $n/2$, we can call L the subgroup of B_n generated by $\{\sigma_1, \dots, \sigma_{m-1}\}$, and U the subgroup generated by $\{\sigma_{m+1}, \dots, \sigma_n\}$. It is clear that L and U commute.

Let us describe the protocol. Three braids p, s, and r are selected. The braid p is public, $s \in L$ is only known to Alice, and $r \in U$ is only known to Bob. Now Alice conjugates p by s and sends it to Bob, and Bob conjugates p by r and sends it to Alice. Now Alice conjugates rpr^{-1} by s, and Bob conjugates sps^{-1} by r. The results are the common key, which are clearly equal as L and U commute.

The security of the protocol is now based on the difficulty of the Diffie-Hellman conjugacy problem (DHCP), which is the following. Suppose we are given elements x, axa^{-1} and bxb^{-1} in B_n such that a and b commute. Then the DHCP asks if it is possible to find $baxa^{-1}b^{-1}$.

7.4.3. Ciphering (Ko et al.) In the next protocol, also defined in [**131**], a computable collision-free hash function $h : B_n \to \{0,1\}^N$, and also *one-way*, i.e., it is not possible to retrieve b from $h(b)$ for $b \in B_n$ is considered. We define L and U as in the previous section, and N is a natural number.

As before, let $p \in B_n$ be a braid, p' a conjugate of p by some elements s. The braids p and p' are public, while $s \in L$ is a private key for Alice. The message m_B to be transmitted is an element of the product $\{0,1\}^N$, and we denote by $+$ the sum, coordinate by coordinate, in $\mathbb{Z}/2$.

Now Bob selects a braid r in U and sends a) the element of $\{0,1\}^N$ given by $m_B + h(rp'r^{-1})$ and b) rpr^{-1}. Now Alice computes $m_B + h(rp'r^{-1}) + h(srpr^{-1}s^{-1})$. As the latter is equal to m_B, Alice retrieves Bob's message.

Again, the security of the problem is based on the Diffie-Hellman conjugacy problem.

7.4.4. Diffie-Hellman for authentication. This protocol is proposed in [**198**], following [**131**]. The starting point are p, s, and p' as in the previous scheme. Here the algorithm is simple. Bob chooses a braid r in U, conjugates p by r and sends the result to Alice. Alice receives it and in turn sends Bob $h(srpr^{-1}s^{-1})$. Then Bob checks if $h(srpr^{-1}s^{-1}) = h(rp'r^{-1})$. If Alice has performed the computation correctly, then the equality holds and Bob accepts. Without the private key, a correct answer to Bob is only possible if the sender has been able to solve the corresponding Diffie-Hellman conjugacy problem. Hence, the security of the scheme relies again in the difficulty of this problem.

7.4.5. Authentication (Sibert et al.) In the mentioned paper [**198**], Sibert et al. propose a new protocol for authentication. The starting point is defining p, s, and the hash function h as in the previous scheme, but not supposing this time that s belongs to a specific subgroup of B_n. In this procedure the next scheme is repeated k times. First, Alice selects a braid $r \in B_n$ and sends $h(rp'r^{-1})$ to Bob. Now Bob chooses a bit c and sends it to Alice. Then, if $c = 0$, Alice sends r, and Bob checks $h(rp'r^{-1}) = h(rpr^{-1})$; if $c = 1$, Alice sends rs, and Bob checks $h(rp'r^{-1}) = h(rsps^{-1}r^{-1})$.

Observe that if Alice sends the correct answer, in both cases she will be accepted. On the other hand, if Alice cheats, the possibility of being accepted is only 0.5. Therefore, after k repetitions this probability decreases to 0.5^k, tending to zero as k grows. For Alice to be accepted with higher probability, she must solve the conjugacy search problem both for (p, p'), so the security of the scheme relies in the complexity of the solution of this problem.

We finish this section by giving some recommendations for the implementation of the schemes.

- For the AAG protocol, it is recommended to work in B_{80}, with $|p_i|$ and $|q_j|$ equal to 5 or 10, and $l = m = 20$.
- For the Diffie-Hellman schemes for Ko et al. authentication, again B_{80} and sequences of length 12.
- For Sibert et al. authentication, it is recommended to use braids in B_{50} and words of length 512.

7.4.6. Discussion on security. The braid-based cryptography, and particularly the previous schemes, have received certain criticism owned to their vulnerability to certain attacks (see [**132**] for an excellent compilation). First, it is clear from the structure of the protocols that a solution of the conjugacy problem and/or their variants for braid groups in polynomial time would seriously compromise the security of the cryptosystems. Although, as mentioned above, no general polynomial solution has been found thus far, it seems not out of reach. Polynomial solutions exist for a number of concrete cases and some variants. In particular, it seems that the average complexity for the construction of Gebhardt's ultra summit set (USS) is polynomial. In particular, using random words, the exact solutions of these problems, and the construction of USS, can be used to attack the previous protocols.

Other length attacks are based on the following idea: given (p, p') as defined in several of the previous protocols, the strategy is to find different conjugates of p' in order to obtain a conjugate that satisfies different minimality conditions. The simpler versions of this idea, which just compare the output with p, have obtained a certain degree of success against key exchange protocols; while a more sophisticated approach, which includes solving a conjugation search problem in the symmetric groups, has shown to be quite effective against the AAG scheme. See [**78**], [**68**], [**109**], and [**140**].

Finally, attacks based on linear representations (as the ones discussed in [**112**] and [**140**]) should be mentioned. In general, the strategy consists, giving p and p' in a braid group B_n and their representations in some linear group G, in solving the conjugacy search problem or some variant inside the linear group, and then to use the information to retrieve a conjugating element in the braid group. Although this method does not work easily for the conjugacy search, in [**31**] it is showed by Cheon-Jun that it can be useful to solve, in some cases, the Diffie-Hellman conjugacy search, and then to break the corresponding cryptosystems.

An interesting survey of different lines of research that attempt to make these attacks useless or at least reduce their impact is given in [**132**]. It is also remarkable that, as an alternative, the affine version of braid groups has been proposed. See [**212**] for more information.

CHAPTER 8

Hyperbolic Groups

8.1. Introduction

Since its introduction by Gromov in his fundamental study of the geometry of the manifolds with nonpositive curvature, hyperbolic groups have played a main role in geometric group theory. Although the properties of these groups are inspired by concepts from hyperbolic geometry, many of their group-theoretic features are well understood; for example, the Tits alternative works for them: a hyperbolic group contains a free group in two generators, or else it is virtually cyclic. Moreover, see our reference below to [**20**].

From the point of view of this book, hyperbolic groups are very interesting because their main decision problems have been thoroughly studied. This feature makes them become very attractive as a source of platforms for cryptographic protocols. Especially important in this context is the concept of distortion and distorted subgroups, which have been shown to be very useful in a cryptographic context.

This chapter is structured as follows. In next section we review the main definitions and examples concerning hyperbolic groups that will be needed in the remainder of the chapter. Taking into account the target of this book, we have tried to be as concise as possible in the geometric properties of the groups. Section 8.3 is devoted to the study of the hyperbolic groups from the point of view of decision problems and algorithmics, while the last section is devoted to the occurrences of these groups in cryptography.

Our main reference for this chapter has been the book of Bridson-Haefliger [**20**].

8.2. Mathematical background

8.2.1. Main definition. We start with the usual definition of the term "hyperbolic group". First we need to recall the definition of a δ-hyperbolic metric space. Recall that a *metric space* is a set X which is endowed with a *distance function* $d : X \times X \to \mathbb{R}$. Moreover, given a subset $A \subseteq X$ and $\delta > 0$, a δ-neighborhood of A is the subset of X given by:

$$A = \{x \in X \mid d(x, a) \leq \delta \text{ for some } a \in A\}.$$

Also, a *geodesic triangle* in a metric space is a triangle whose sides are geodesic lines. Moreover, a metric space is said to be *geodesic* if there is always a geodesic between two points. Now we can define the basic concept of the *δ-slim triangle.*

DEFINITION 8.1. *A geodesic triangle in a metric space is said to be* δ-slim, *for some $\delta > 0$, if each of its sides is contained in a δ-neighborhood of the union of the remaining two.*

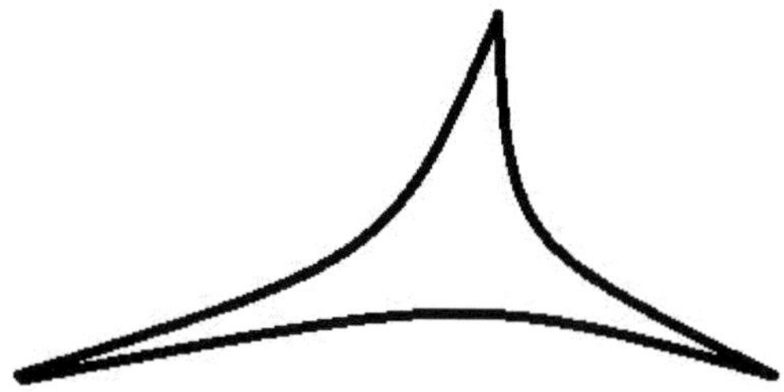

FIGURE 8.1. A δ-slim triangle.

The existence of the δ-slim triangle characterizes the hyperbolic spaces.

DEFINITION 8.2. *For $\delta \geq 0$, a geodesic metric space X is said to be* δ-hyperbolic *if every geodesic triangle is δ-slim. Moreover, the space is said to be* hyperbolic *if it is δ-hyperbolic for some $\delta > 0$.*

Here, δ gives a degree of hyperbolicity.

Now given a connected graph $\Gamma = (V, E)$, we can always define the *graph metric* over it in the following way. First, a length function $\lambda : E \to (0, \infty)$ is defined. For every $e = vw$, consider an arc f_e such that $f_e(0) = v$ and $f_e(1) = w$. Then, given a partition $0 = t_0 < t_1 < \ldots < t_n$ of $[0,1]$, a piecewise linear path $c : [0,1] \to \Gamma$ is a map such that the restriction to every subinterval $[t_i, t_{i+1}]$ is a composition $f_{e_i} \circ c_i$, with e_i being an edge of the graph and $c_i : [t_i, t_{i+1}] \to [0,1]$ an affine map. Now the length of c is defined as the sum

$$l(c) = \sum_{i=0}^{n-1} \lambda(e_i)|c_i(t_i) - c_{i+1}t_{i+1}|,$$

and the distance between two points x and y is defined as the infimum of the lengths of the piecewise linear paths that start at x and end at y. In our context, the length function takes the value 1 in every edge.

We also need to recall the definition of the *Cayley graph*. Let G be a group, S a system of generators. It is said that Γ is the Cayley graph of G with respect to S if there is a vertex of Γ for every element of G, and there is an edge between g and h such that there exists $s \in S$ such that $gs = h$.

Now we are ready to state the definition of a *δ-hyperbolic group*.

DEFINITION 8.3. *For $\delta > 0$, a group G is called* δ-hyperbolic *if for some presentation of the group, the Cayley graph, endowed with the graph metric, is a δ-hyperbolic space.*

A group which is δ-hyperbolic for some $\delta > 0$ is called *hyperbolic* or *word-hyperbolic.*

Observe that different systems of generators may give nonisomorphic Cayley graphs, but the previous definition does not depend on the choice of such a system.

The hyperbolic groups admit different definitions and characterizations; for example, in terms of the word metric or using an isoperimetric inequality.

8.2.2. Examples of hyperbolic groups. It is very important in our context to identify many examples of hyperbolic groups, as some of the algorithmic results of the next section can be applied to the whole class. We will soon review different families of groups that are known to be hyperbolic. Before, we recall some concepts from group theory and topology that can be useful to understand the examples.

8.2.2.1. *Virtually cyclic groups.* Recall from the first chapter that a group G is virtually cyclic if it possesses a cyclic subgroup of finite index. Clearly this class contains finite groups and extensions of $\mathbb{Z}$ by finite groups. In the context of hyperbolicity, virtually cyclic groups are usually called elementary hyperbolic groups.

8.2.2.2. *Surface groups.* From the classification theorem of connected closed surfaces, it is deduced that the nontrivial fundamental groups of this surfaces can only be:

- $\langle a_1, b_1, \ldots, a_n, b_n \mid a_1 b_1 a_1^{-1} b_n^{-1} \ldots a_n b_n a_n^{-1} b_n^{-1} \rangle$,
- $\langle a_1, \ldots, a_n \mid a_1^2 \ldots a_n^2 \rangle$.

The first case corresponds to the orientable case, while the second corresponds to the nonorientable. These groups are part of the bigger class of one-relator groups, defined in the first chapter, and of the family a fuchsian groups, that we define next.

8.2.2.3. *Fuchsian groups.* Fuchsian groups are examples of groups with small cancellation conditions that will treated with detail in a separate chapter.

DEFINITION 8.4. *A group G is said to be* fuchsian *if it admits a presentation with generators $\{a_1, b_1, \ldots, a_g, b_g, x_1, \ldots x_n, f_i, f_k\}$, for $g \geq 1$, $n, k \leq 0$, and relations $x_1^{m_1} = \ldots = x_n^{m_n} = 1, x_1 \ldots x_n f_1 \ldots f_k \prod_{i_1}^{g} [a_i, b_i]$. Here the m_i are finite natural numbers.*

8.2.2.4. *Artin-Tits groups.* Consider a group G with a finite number of generators $\{a_1, \ldots, a_n\}$. For every distinct pair (i, j), $i, j \leq n$, let $m_{ij} \leq 2$ be a natural number or ∞. We denote the word $a_i a_j a_i a_j \ldots$ with m_{ij} letters by $\langle a_i a_j \rangle^{m_{ij}}$ if m_{ij} is natural, and the empty word otherwise.

DEFINITION 8.5. *A group G with generators $\{a_i\}$ and only relations of the form $\langle a_i a_j \rangle^{m_{ij}} = \langle a_j a_i \rangle^{m_{ji}}$ with $m_{ij} = m_{ji}$ is called an* Artin-Tits *group. The generators $\{a_i\}$ are called the Artin generators.*

The graph groups that will be studied in Chapter 11 are in particular Artin-Tits groups with $m_{ij} = 2$. Also the braid groups that we reviewed in Chapter 7 are as well with $m_{i,j} = 2$ if $|i - j| > 2$, and $m_{i,j} = 3$ otherwise.

8.2.2.5. *Coxeter groups.* In particular we are interested here in Coxeter groups, which are canonical quotients of Artin-Tits groups.

DEFINITION 8.6. *A group G is said a* Coxeter group *if there exists an Artin-Tits group A with a system $\{a_i\}$ of Artin generators such that G is the quotient of A by the normal subgroup generated by all the relations $a_i^2 = 1$.*

8.2.2.6. *$CAT(k)$-spaces.* Consider the unique 1-connected complete manifold M_k of constant curvature k. For example, M_1 is the sphere, M_0 is the plane, and M_{-1} is the hyperbolic plane.

DEFINITION 8.7. *Let* (X, d) *be a geodesic metric space and*

$$T = \Delta([p, q], [q, r], [p, r])$$

be a geodesic triangle in X. *Then for fixed* k, *a triangle* $T' = \Delta([p', q'], [q', r'], [p', r'])$ *in* M_k *is called a* comparison triangle *for* T *if there are equalities of lengths* $d(p, q) = d(p', q')$, $d(q, r) = d(q', r')$ *and* $d(p, r) = d(p', r')$. *Given a point* x *in a side of the triangle* T, *say* $[p, q]$, *its* comparison point x' *in* T' *is the point in* $[p', q']$ *such that* $d(p, x) = d(p', x')$. *Comparison points for points in the interior of* T *are defined by linearity.*

Under the previous conditions, it is said that the CAT(k)-inequality *holds for* T *if for every* x, y *in* T, *the distance* $d(x', y')$ *between their comparison points is bigger than* $d(x, y)$.

Now we can define $CAT(k)$-spaces.

DEFINITION 8.8. *With the previous notation, a space* X *is called a* $CAT(k)$-space, $k \leq 0$, *if the* $CAT(k)$*-inequality holds for all the triangles in* X. *If* $k > 0$, *the condition to be* $CAT(k)$ *is only required for triangles with perimeter smaller than* D_k, *being* $D_k = \pi/\sqrt{k}$ *the diameter of* M_k.

We are interested in groups that act in these spaces.

8.2.2.7. *Hyperbolic manifolds.* Recall that a closed manifold is *hyperbolic* if it is the quotient of a hyperbolic space by a torsion-free subgroup of isometries. Examples of hyperbolic 3-manifolds are the hyperbolic space H^3 and the Seifert-Weber dodecahedral. The classical technique of Dehn filling is an standard method to obtain hyperbolic manifolds with exotic fundamental groups.

8.2.2.8. *Baumslag-Solitar groups.* A very common source of examples and counterexamples in the context of geometric group theory are the Baumslag-Solitar groups, that we define next.

DEFINITION 8.9. *The* Baumslag-Solitar group *is defined, for* $m, n \in \mathbb{Z}$, *by the presentation:*

$$\langle a, b \mid a^n = ba^m b^{-1} \rangle.$$

8.2.2.9. *Families of hyperbolic groups.* Now that we have given the appropriate definitions, we may list different families of hyperbolic groups:

- Virtually cyclic groups, i.e., groups with an infinite subgroup of finite index. Clearly this class contains the finite groups. In this context, these groups are usually called elementary hyperbolic groups. See [**92**, Section 3.2].
- Free groups, or more generally, groups that act on trees with finite stabilizers. A classic example of the latter is $SL(2, \mathbb{Z})$. See [**92**, Proposition 3.2.A].
- Fundamental groups of nonorientable compact surfaces and other Fuchsian groups are also hyperbolic groups. See [**92**, Section 3.3], [**20**, III.Γ, Theorem 2.6 and references therein].
- One-relator groups that contain torsion. This is a consequence of [**167**, Theorem 3].
- Coxeter groups that do not admit subgroups isomorphic to the free abelian group in two generators. See [**157**, Theorem 17.1].

- Groups that satisfy the small cancellation conditions $C'(1/4) - T(4)$ or $C(6)$. In particular, fundamental groups of orientable surfaces of genus bigger than two. See [**75**, Theorem 33].
- Groups that act geometrically on $CAT(k)$-spaces, $k < 0$. See for example [**124**, Section 2].
- The fundamental groups of hyperbolic manifolds. The classical technique of Dehn filling is an standard method to obtain exotic hyperbolic groups. See [**75**, Chapter 3].
- Some semidirect products of fundamental groups of closed manifolds with an infinite cyclic groups.

It is also important to mention examples of groups which are *not* hyperbolic. In this sense, a key fact is that no hyperbolic subgroup can contain free abelian subgroup of rank 2 ([**20**, III.Γ, 3.10]). Moreover, no group that contains the Baumslag-Solitar group $B(m,n)$ is hyperbolic. See ([**74**, Corollary 6.8]), and take into account that $B(m,m)$ always contains $B(1,1) = \mathbb{Z}^2$ for $m \geq 1$.

8.3. Algorithmic problems

Now we will describe different approaches that have been proposed for several decisions problems in the wide context of hyperbolic groups. The algorithmic approach will be crucial for the cryptosystems defined in next section. We continue following [**20**].

8.3.1. The word problem. The word problem is solvable in hyperbolic groups by means of the classical *Dehn algorithm*. Before we recall it, we will need the following definition:

DEFINITION 8.10. *Let G be a group, $\langle S \mid R\rangle$ a finite presentation. Assume that there exist finite sets of words $\{u_i\}$ and $\{v_i\}$, $1 \leq i \leq n$, such that:*

- *For every i, u_i and v_i represent the same element of G.*
- *For every i, the length $|v_i|_S$ of v_i in the generators of S is smaller than the length of u_i.*
- *Every nontrivial word that represents the trivial element of G contains u_i for some $i \leq n$.*

Then if $R = \{u_1v_1^{-1}, \dots, u_nv_n^{-1}\}$,$\langle S \mid R\rangle$ is called a Dehn presentation *for G.*

A Dehn presentation always provides a method to solve the word problem:

PROPOSITION 8.11 ([**20**], III.Γ, 2.4). If G is a group that possesses a Dehn presentation, the word problem is solvable for it.

PROOF. The proof is simple. First take a word w, and check if it contains a word u_i, for some $i \leq n$. If it does not, then the algorithm stops and w cannot represent the identity. If it does, then the subword u_i is replaced by v_i, which strictly diminishes the length of w. Iterating the scheme, we will arrive at the trivial word in a finite number of steps if w represents the identity, or the algorithm stops before if w does not. □

The following theorem of Gromov implies that the word problem is solvable for hyperbolic groups. In fact, it is a characterization of hyperbolic groups.

THEOREM 8.12 ([**20**], III.Γ, 2.6). *A group is hyperbolic if and only if it has a Dehn presentation.*

The solution of the word problem using the Dehn algorithm is very efficient, as is shown by the next result.

PROPOSITION 8.13 ([**44**], Section 3). Given a Dehn presentation of a group G, the Dehn algorithm solves the word problem in linear time in a multitape Turing machine.

8.3.2. The conjugacy problem. As is the case with the word problem, the conjugacy problem is efficiently solvable in hyperbolic groups.

Let us first see how to obtain an exponential algorithm for the conjugacy problem, which is a consequence of the quasi-monotone conjugacy property. We recall now the definition:

DEFINITION 8.14. *Let $G = \langle S \mid R\rangle$ be a group with S finite. Suppose that given any pair of conjugate words u and a positive constant K, there always exists a word $w = s_1 \dots s_n$ in the generators of S such that if we denote $w_i = s_1 \dots s_i$, we have $w^{-1}uw = v$ and $d(1, w_i^{-1}uw_i) \le K max(|u|, |v|)$, for $i \le n$. Then we will say that G possesses the* quasi-monotone conjugacy property.

Of course, here d is the distance in the Cayley graph associated with the presentation, and $|-|$ is the length of the word in the generators of S.

We give two approaches to the solution of the conjugacy problem. Assume now that the word problem is solvable for a group G that possesses the quasi-monotone conjugacy property, and a system of generators and relations in G is fixed. Given two words u and v of length $\le n$ for a certain fixed n, the solvability of the word problem implies that it is possible to decide if u and v are conjugate.

Now consider a graph $\Gamma(n)$ whose set of vertices is given by the words of G with length $\le n$, and such that there is an edge between u and v if u is conjugate to v. By the previous paragraph, there is an algorithm to construct $\Gamma(n)$ for every n. Now by the quasi-monotone conjugacy property, u and v are conjugate if and only if they lie in the same path-connected component of $\Gamma(K\max(|u|, |v|))$. Thus, we have a general procedure to determine if u and v are conjugate.

Using the fact that the number of vertices of the graph $\Gamma(n)$ is bounded by $2|S|^n$, it can be shown that there exists a word that conjugates u to v with length that is exponentially bounded by $\max(|u|, |v|)$. As we saw in the previous section hyperbolic groups have solvable word problem in linear time, we obtain that the algorithm we have described solves the conjugacy problem for these groups in exponential time.

It is also possible to solve the conjugacy problem for hyperbolic groups in linear time, although the algorithm is less transparent. In the algorithm, which we describe below, the role of the hyperbolicity is stated in the following lemma:

LEMMA 8.15 ([**20**], III.Γ, Section 2, see also [**4**], Lemma 3.1). *Consider a δ-hyperbolic group $G = \langle S \mid R\rangle$, with S finite. There exists a constant $K(\delta) > 0$ such that if u and v are words representing conjugate elements of the group G, and all the subwords of u and v and their cyclic permutations of length $\le 8\delta + 1$ are* *geodesic, then one of the two following alternatives hold:*

- *The maximum of the lengths of u and v is $\leq 8\delta + 1$.*
- *There is a word w such that $|w| \leq 2\delta + 1$, and cyclic permutations u' and v' of u and v, such that w conjugates u' to v'.*

Now we can sketch this solution of the conjugacy problem. Let G be given by the notation of the previous lemma, with $\langle S \mid R \rangle$ being a Dehn presentation. We should decide if two words u and v represent conjugate elements. First search inside u and v and their cyclic permutations for all the subwords of length $\leq 8\delta+1$ that are not geodesic, and replace them with geodesic words. After all these replacements, which can be undertaken in linear time in $|u| + |v|$, we have obtained words u' and v' which are respectively conjugate to u and v and such that all their subwords and the subwords of their cyclic permutations of length $\leq 8\delta + 1$ are geodesic. Now by the previous lemma, we obtain a finite set $W = \{w_1, \ldots, w_n\}$ of words such that u' is conjugate to v' if and only if some element of W conjugates u' to v'. This set W can be taken as the set of words of length $\leq K$, joined with one element for each pair of conjugate elements whose length is $\leq K$. Now applying Dehn algorithm, and taking into account that the set W is finite, we may decide if w_i conjugates u' to v' for some $i \leq n$.

In [**52**], an even more efficient algorithm has been proposed.

We finish this section by mentioning that Bridson-Howie show that the conjugacy search problem for lists is solved in quadratic time for hyperbolic groups. This fact makes these groups a noninteresting platform for public-key encryption systems based on this problem, a question that generated some interest in the past. More information can be found in [**21**].

8.3.3. The isomorphism problem. The isomorphism problem for torsion-free hyperbolic groups was solved by Sela in the important paper [**192**]:

THEOREM 8.16. *Let G_1 and G_2 be two torsion-free hyperbolic groups. Then there exists a procedure that decides in finite time if the groups are isomorphic.*

The construction of the algorithm is complicated and highly technical, but a little sketch of it can be given here. The argument is based on the concept of the *stretching factor*, which we now define.

DEFINITION 8.17. *Given a homomorphism $h : G_1 \to G_2$ between two groups, the* stretching factor *$\sigma(h)$ is defined as the minimum of the numbers*

$$max_{x \in G_1} |y^{-1} h(x) y|$$

for all the elements $y \in G_2$.

It has been proved that all the stretching factors for the monomorphisms $G_1 \to G_2$ can be uniformly bounded, and that a bound σ can be computed in an effective way. Moreover, conjugating homomorphisms do not change the stretching factor. Given a number N, only *finite* conjugacy classes of homomorphisms $G_1 \to G_2$ have N as a stretching factor, and these homomorphisms can be enumerated. As the uniform bound is an integer, we can list all the homomorphisms whose stretching factor is $\leq \sigma$. After performing the same process for homomorphisms $G_2 \to G_1$, it is enough to check the corresponding compositions in order to decide if G_1 and G_2 are isomorphic.

8.3.4. Other algorithmic problems.

Membership problem. Let us begin by recalling the definition.

DEFINITION 8.18. *Given a geodesic metric space (X, d), a subspace $Y \subseteq X$ is* quasiconvex *if there exists a constant $K \geq 0$ such that for every pair of points of Y and for every x that belongs to a geodesic that joins the two points, $d(x, Y) \leq K$. A subgroup $H < G$ is quasiconvex if the Cayley graph of H is quasiconvex in the Cayley graph of G.*

These groups behave well in the following context.

PROPOSITION 8.19. The membership problem is solvable for quasiconvex subgroups of hyperbolic groups.

There are different proofs of this fact available (see for example [**76**], [**54**], or [**123**]). On the other hand, it can be proved by means of the Rips construction [**184**] that there are examples of nonquasiconvex subgroups of hyperbolic groups for which the membership problem is unsolvable. This is obtained by means of small cancellation properties, which we will define in Chapter 9.

On the other hand, in their work [**30**], Chatterji-Kahrobaei-Lu proved that the membership problem is solvable in polynomial time for the free-by-cyclic hyperbolic groups constructed by Mecham-Mukherjee in [**153**]. These groups will be useful in cryptographic platforms, as we will see in next section. Another solvable membership problem is used in the Flores-Kahrobaei-Koberda protocol of Section 8.4, under hypotheses that will be established there.

Geodesic problems. For hyperbolic groups, the geodesic problem, and hence the geodesic length problem, have an efficient solution:

PROPOSITION 8.20. *If G is a hyperbolic group, its geodesic problem is solvable in polynomial time.*

PROOF. See Chapter 11 in [**51**] and [**30**, Theorem 2.4]. □

8.4. Cryptographic protocols

In this section we will review some cryptographic protocols that have been proposed for which hyperbolic groups work as a convenient platform.

We start by recalling recent work of Chatterji-Kahrobaei-Lu [**30**]. There, they propose two different versions of two protocols, a basic one and a secure one. We will describe the basic versions next, and show how to modify them in order to make them secure.

Protocol 1. Here we suppose that Alice wants to send a message to Bob through a channel that is not secure. It is assumed that a finitely generated group $G = \langle S \mid \mathrm{R} \rangle$ is known publicly, and a subgroup $H < G$ generated by a system $T = \{t_1, \dots, t_s\}$ is only known to Alice and Bob and is distorted in G. It is required that the geodesic length problem and the membership problem are solvable in polynomial time for G and H.

Under these conditions, Alice picks a word $h \in H$ whose geodesic length $l_T(h)$ in T is n, then writes it in a geodesic way in terms of S (so $l_S(h) \leq n$) and sends it to Bob. Now Bob rewrites h in terms of T and obtains n, which is the message.

A possible vulnerability of this scheme comes from the fact that if an eavesdropper intercepts many elements of H, it could be possible that he/she generates H

using these elements, and then be able of reconstruct the message h. To avoid this, the authors propose a modification of the protocol, in which Alice sends, together with the elements of H that codify the message, random words in the generators of G that make impossible for the eavesdropper to identify the distorted subgroup. For this modified protocol to be feasible, the subgroup membership problem should be efficiently solvable for the group. Moreover, Alice and Bob must secretly share a generator of (pseudo)random numbers and a seed for it.

Protocol 2. Different from the previous protocol, in this one the "big" group $G = \langle S \mid \rangle$ is secret, while the distorted subgroup H generated by T is public. We assume again that the geodesic length problem is efficiently solvable for G.

Alice picks an element $g \in H$ such that $|g|_G = n$. Again this is the message that should be sent. She expresses the message in terms of T, so $|g|_H = m \geq n$, and sends that expression of this element g to Bob. Then Bob expresses again g in terms of S and recovers n.

Now a possible problem is that an eavesdropper can determine approximations of the different messages by testing upper bounds of them. Again the authors present a modification of the protocol by introducing another group K in which H can be embedded as an exponentially distorted subgroup. This group is secretly known for Alice and Bob. Now the element g is sent in terms of generators of K, and the exponential distortion makes any upper bound of n too far from the "correct" message n. As a bonus, as g is sent in terms of generators of K, an eavesdropper has no way to guess G.

For the first protocol, the free-by-cyclic hyperbolic groups of Mechelen-Mukherjee quoted at the end of last section are an appropriate platform. For the second, an explicit example is described in [**30**, 4.2], based on results of [**7**].

In [**62**], low-dimensional topology tools are applied to define a new public-key cryptographic protocol. Although in general we are introducing for each class of groups and protocols the necessary mathematical background to make every chapter self-contained, we will make an exception here, as the concepts needed to develop this protocol are beyond the scope of this book. We will simply describe the protocol, referring the interested reader to the paper mentioned for the needed definitions and results.

We start by listing the public information:

- A hyperbolic 3-manifold M fibered over the circle, such that the rank of its first real cohomology group is bigger than one.
- A finite presentation of the fundamental group of the manifold.
- A fibered face F of the Thurston normed ball.
- For every primitive integral cohomology class $\phi \in \mathbb{R}_+ F$, a surface presentation P_ϕ of the fundamental group of the associated surface S_ϕ, an automorphism ψ_ϕ of this group, and a letter $t_\phi \in \pi_1 M$ such that π_M admits the presentation:
$$\{P_\phi, t_\phi \mid t_\phi x t_\phi^{-1} = \psi_{\psi_\phi(x)}\},$$
where $x \in \pi_1(S_\phi)$.
- An explicit isomorphism between the previous presentation of $\pi_1(S_\phi) \rtimes \mathbb{Z}$ and $\pi_1(M)$.
- A function efficiently computable $G \to \mathrm{Prim}(\mathbb{R}_+ F)$, where G is a finitely generated group and $\mathrm{Prim}(\mathbb{R}_+ F)$ is the set of primitive integral cohomology classes.

Note that $\mathrm{Prim}(\mathbb{R}_+F)$ is an infinite set. Instead, in practice we will not consider the whole collection, but a collection large enough in order for Alice and Bob to make their choices.

Now we can describe the protocol. First, by means of any public-key cryptography scheme, Alice and Bob select a private key $g \in G$. Both apply the function f to g, obtaining in this way the corresponding cohomology class, the group $\pi_1(S)$, the presentation P and the automorphism ψ (here we suppress the subindex $f(g)$ for simplicity). For every generator in P, and given a fixed positive integer N, Alice computes the geodesic length. The maximum of these lengths, say $|s_{max}|$, is reached in a generator s_{max}. This will be the key to be transmitted. Now Alice sends a collection C of elements of $\pi_1(M)$, such that written in terms of the chosen generators their lengths are similar to N. She also imposes the condition that for every generator s of $\pi_1(S)$, $\phi^N(s)$ belongs to C, and no other element of C belongs to $\pi_1(S)$. Then Bob checks which elements of C belong to $\pi_1(S)$, recover N conjugating by an element of $\pi_1(M)$, and finally gets $|s_{max}|$ by evaluating ψ^M over the generators of $\pi_1(S)$.

The feasibility of the scheme comes from the fact that under these conditions, $\pi_1(S)$ is exponentially distorted in $\pi_1(M)$, and from the following result, stated as [**61**, Proposition 5.4]:

PROPOSITION 8.21. *Under the previous conditions, the membership problem for $\pi_1(S)$ and $\pi_1(M)$ is solvable in linear time.*

Some final remarks to this protocol:

- In [**62**, Section 4], a concrete example of application of the previous protocol is carried out. The manifold fits in a fibration $S \to M \to S^1$, where S is the three-punctured disk.
- A simplified version of the described protocol gives rise to a symmetric-key exchange protocol. See Section 5 of the mentioned paper.
- The described protocol has several levels of security. In the first level, its security is similar to that of the classical Anshel-Anshel-Godfeld scheme [**2**]. Second, the interception of N does not imply the interception of the key, as the latter is exponentially bigger than the former. Finally, an attack to the database of cohomology classes without knowledge of the secret fiber can be made inconsequential by making the database much larger than the size of the key.

CHAPTER 9

Small Cancellation Groups

9.1. Introduction

In this chapter we will review some applications of small cancellation theory to cryptography. Roughly speaking, a presentation of a group shows small cancellation conditions if the words that define the relators only overlap in a few letters. The basic notions go back to Dehn and Tartakowski, who generalized surface groups in this way, and the theory was fixed the after work of Lyndon and Schupp in the sixties. Recent advances in small cancellation theory have been important in the theory of infinite torsion groups and groups of intermediate growth. Many of the ideas that arose in relation with these concepts have been important in developing the corpus of hyperbolic groups. The general reference used here is Chapter 5 of [**150**].

This chapter is structured in the following way. First we will introduce the main notions of cancellation theory that will be needed later. As in the rest of the book, we provide only the necessary information, referring the interested reader to other sources which are not directly related with cryptographic topics. Then we will consider small cancellation groups from the algorithmic point of view. We conclude the chapter by reviewing two protocols that propose using small cancellation groups as platforms groups. First, we describe a key exchange protocol, the commuting action key exchange, defined by Shpilrain-Zapata in [**197**] and state why small-cancellation non-hyperbolic groups are an adequate platform in this context. Then, we will introduce a variation of the Habeeb-Kahrobaei-Shpilrain secret sharing scheme [**97**] using a shortlex order developed by Cavallo-Kahrobaei [**27**], and why $C'(1/6)$-small cancellation groups (see definition in next section) behave well in this context.

In this chapter we mainly follow the exposition of Schupp's survey [**190**].

9.2. Mathematical background

We will begin with some definitions.

DEFINITION 9.1. *Let F be a free group generated by $S = \{x_1, \dots, x_n\}$. A* letter *in F is an element of S or an inverse of an element of S.*

DEFINITION 9.2. *Let $w = y_1 \dots y_m$ be a reduced word in the letters of S. Then w is said to be* cyclically reduced *if $y_1 \neq y_m^{-1}$. A subset $R \subseteq F$ is called* symmetrized *if the following conditions hold:*

- *Every element of R is cyclically reduced.*
- *Every cyclically reduced conjugate of an element of R is also in R.*

In the following we will define what is understood as "small cancellation". First we need to introduce the concept of a piece.

Definition 9.3. *Let R be a symmetrized subset of a free group. A* piece *in R is a common prefix of at least two words in R, i.e., a word a such that there exists w_1 and w_2 in R with $w_1 = ab_1$ and $w_2 = ab_2$.*

Observe that in the terms of this definition, the piece a is cancelled via the product $w_2^{-1}w_1$. In general, small cancellation will refer to symmetrized subsets where the pieces are not very big. We will now state the main small cancellation conditions.

Definition 9.4. *Let R be a symmetrized subset of a free group with a fixed finite number of generators.*

- *Condition $C'(\lambda)$. If a piece $a \in R$ is a prefix of a word w, then $|a| < \lambda|w|$.*
- *Condition $C(p)$. There are no elements in R that can be written as a product of less than p pieces.*
- *Condition $T(q)$. Let $h \in [3, q)$ be an integer. Consider h elements $\{r_i\}$ in R such that r_i and r_{i+1} are not inverse to each other for any integer $i \in [1, h-1]$, and consider all the pairs $r_1r_2, r_2r_3, \ldots, r_{h-1}r_h, r_hr_1$. Then at least one of these words is reduced. This is usually called the* triangle condition.

Observe that the first condition is a *metric condition*, as it depends on the length of the words in the free group, while the other two are not. It can be seen that $C'(\lambda)$ always implies $C(p)$ for $\lambda \leq 1/(p-1)$.

Let $G = \langle S \mid R \rangle$ be a group. Assume that R is a symmetrized subset in the free group $F(S)$. Then we will say that G satisfies a small cancellation condition if R satisfies the condition with respect to $F(S)$.

It is possible to extend the definition of cancellation conditions for every finitely generated group, in the following way.

Definition 9.5. *Let F a free group and T a subset of F. The* symmetrized closure *of T in F, denoted $R(T)$, is the subset of F whose elements are the cyclic permutations of the words in T and of their inverses.*

A finitely generated group $G = \langle S \mid T \rangle$ is said to satisfy a small cancellation condition if $R(T)$ satisfies the condition with respect to $F(S)$.

Now we will review some examples, following the exposition of [**201**].

Example 9.6. The groups that motivated the development of the theory were the fundamental groups of closed connected orientable surfaces, which for genus $g \geq 1$ have the presentation:

$$\langle a_1, \ldots, a_g, b_1, \ldots, b_g \mid a_1b_1a_1^{-1}b_1^{-1} \ldots a_gb_ga_g^{-1}b_g^{-1} \rangle.$$

Applying the definitions directly to the symmetrized version of the relators, it can be seen that this group satisfies the small cancellation conditions $C'(1/(4g-1))$ and $C(4g))$. Observe that the only pieces are the trivial element and the generators and their inverses; hence, the maximum length of a piece is one. It can also be checked that the condition $T(4g)$ holds.

Example 9.7. Now consider the fundamental groups of closed connected nonorientable surfaces of genus g given by the presentation:

$$\langle a_1, \ldots, a_g \mid a_1^2a_2^2 \ldots a_g^2 \rangle.$$

By similar reasons to the above example, the maximum length of a piece is one, and the group satisfies the conditions $C'(1/(2g-1))$, $C(2g)$ and $T(2g)$.

EXAMPLE 9.8. An important family of one-relator groups is given by the following presentation:

$$\langle a_1, \ldots, a_n \mid z^m \rangle,$$

where $m > 1$ and z involves all generators and is not a proper power in the group. As some generators should appear twice in the expression of z^m, every piece in the symmetrization of R should be a proper subword of z or of its inverse. In particular, $C'(1/m)$ and $C(m+1)$ hold. The condition $C(m+1)$ can be improved to $C(2m)$ (see [**179**]), but in general $C'(1/(m+1))$ does not hold.

EXAMPLE 9.9. On the other hand, consider the finite cyclic group

$$\langle a \mid a^m \rangle,$$

with $m \geq 1$.

As the cyclic permutations do not alter a^m nor a^{-m}, the only possible piece is the empty piece. Hence, all the small cancellation conditions hold for the group.

It is also worth mentioning that given a free product $G_1 * G_2$ of two groups G_1 and G_2 and a small cancellation condition C, the weaker value between C applied to G_1 and to G_2 holds for $G_1 * G_2$.

9.3. Algorithmic problems

For the protocols that we describe in next section it will be very important to understand the solvability of two Dehn problems in small cancellation groups: the word problem and the conjugacy problem. Hence, in this section we will give an extensive overview of this topic. We follow Section 2 of [**190**], and we refer the reader there for details.

9.3.1. The word problem. Recall that in Chapter 8 the Dehn algorithm was described. The key observation that Dehn formulated for the primary example of a surface group is that if a word is trivial in this group, then the word should contain at least a half of a cyclic permutation of the relator or its inverse.

The power of the Dehn algorithm in the small cancellation groups is a consequence of Greendlinger's lemma, which we now state.

THEOREM 9.10. *Consider a group $G = \langle F \mid R \rangle$, where $R \subset F$ is a symmetrized subset and N its normal closure. Assume that some of the following three pairs of conditions hold for G:*

(a) *$C(6)$ and $T(3)$,*
(b) *$C(4)$ and $T(4)$,*
(c) *$C(3)$ and $T(6)$.*

Then for every nontrivial word w in N, there is a cyclically reduced conjugate w^* of w such that $w^* = u_1 s_1 \ldots u_m s_m$, and such that for every j, there exists a word $r_j \in R$ such that $r_j = s_j b_1 \ldots b_{i(s_k)}$, where the b_i's are pieces. Moreover, m and the numbers $i(s_k)$ should satisfy

- $[\sum_{i=1}^{m}(4 - i(s_k))] \geq 6,$
- $[\sum_{i=1}^{m}(3 - i(s_k))] \geq 4,$
- $[\sum_{i=1}^{m}(5/2 - i(s_k))] \geq 6$

if conditions (a), (b), *or* (c) *hold for G, respectively.*

PROOF. See Section 4 of [**82**] and Section V.4 of [**150**]. □

Observe that, if the cancellation condition $C'(1/6)$ holds for the groups, then so does the first pair of conditions of the statement, as $C'(1/6)$ implies $C(6)$ and the condition $T(3)$ always holds. Then, the groups $C'(1/6)$ are under the hypotheses of the lemma, and it is easy to see that this implies that for these groups, w contains more than half of the elements of R. Then Dehn's algorithm can be applied, and it follows that:

COROLLARY 9.11. *The word problem is solvable for groups for which the $C'(1/6)$ condition holds.*

PROOF. It follows from the previous discussion and [**150**, Theorem V.6.3]. □

In fact, these groups are hyperbolic, so this is a particular case of the solution of the word problem for hyperbolic groups.

The analogous result for the conditions $C(4)$ and $T(4)$ is true, but it is not directly deduced from Greedlinger's lemma.

PROPOSITION 9.12. *The word problem is solvable for groups for which one of the three previous pairs of conditions hold.*

PROOF. The proof relies on a combinatorial area formula, which was proved by Lyndon. See Section V.6 of [**150**] for details. □

It has been proved that the Dehn algorithm solves the word problem in small cancellation groups in polynomial time. The proof is a consequence of results in [**151**]. In particular, for groups for which conditions $C(4)$ and $T(4)$ hold, the problem is solvable in quadratic time; see [**150**, V.6.3].

9.3.2. The conjugacy problem. In small cancellation groups, the conjugacy problem is solvable by means of a variation of Dehn algorithm. We will review this next in groups for which the cancellation conditions of the previous section hold. Again the context is a group $G = \langle S \mid R \rangle$, where R is a symmetrized subset of S, and N is the normal closure. Assume that the word problem is solvable for G.

Consider two words v and w. We would like to check whether they are conjugate or not. We may assume that v and w are cyclically reduced (otherwise we can effectively reduce them to cyclically reduced representatives of the same elements). Moreover, according to the previous section, we may also assume that no cyclic permutation of these words contains half of an element of R (otherwise it would be trivial). Hence, we may assume that u and w are not trivial.

We know that v and w are conjugate in G if and only if some cyclic permutations v^* and w^* are conjugated via a subword u of an element $r \in R$. As there is only a finite number of such subwords u and the word problem is solvable in the group, it is enough to check case-by-case if any one of them gives the desired conjugation. In fact, the length of u is bound by the length of r divided by eight. Hence we have proved:

PROPOSITION 9.13 ([**150**], Theorem V.7.6). *Let G be a group for which one of the following pairs of cancellation conditions hold:*

(a) $C(6)$ *and* $T(3)$,
(b) $C(4)$ *and* $T(4)$,
(c) $C(3)$ *and* $T(6)$.

Then the conjugacy problem is solvable for G. In particular, it is solvable if $C'(1/6)$ holds for G.

Recall from Chapter 8 that the conjugacy search problem is efficiently solvable in hyperbolic groups, and hence in $C'(1/6)$-groups. In particular, by the work of Domanski–Anshel [**44**], there is a multitape Turing machine that solves the word problem in linear time for these groups.

Although a polynomial algorithm is known for the conjugacy problem in finitely generated groups with conditions $C(4)$ and $T(4)$ [**150**], this is not the case in general for the conjugacy search problem in groups for which $C'(\frac{1}{6})$ also holds. This makes these groups useful as platforms for the Shpilrain-Zapata protocol, as we shall see in the following section.

9.4. Cryptographic protocols

In this section we will review two cryptographic protocols for which small cancellation groups have been proposed as platforms. The first is a secret sharing scheme, while the second is a key exchange protocol. We will discuss why groups in this family are appropriate to implement these schemes.

9.4.1. Secret sharing scheme. We have already described the Habeeb-Kahrobaei-Shpilrain secret sharing scheme (HKS), as well as its modification as a (k,n)-threshold scheme following ideas of the classical Shamir's sharing scheme. This modification, in fact, can be as an adaptation of Shamir's protocol to the context of group-based cryptography, which in turn gives rise to some advantages, as a more rational distribution of the shares when the protocol is applied repeatedly. See Chapter 4 for details about the protocol.

Next we follow work of Cavallo-Kahrobaei [**27**]. There, these authors propose using small cancellation groups as an ideal platform to implement the (k,n)-threshold HKS scheme. Specifically, they suggest the groups for which the condition $C'(1/6)$ holds. The reasons of their proposal are twofold.

First, it was discussed above that every group that serves as a potential platform for the HKS scheme must have an efficiently solvable word problem in order for the participants to determine the triviality or not of the corresponding words in the algorithm. As mentioned in the previous section, Domanski-Anshel proved that the Dehn algorithm solves the word problem with linear complexity on groups for which condition $C'(1/6)$ holds.

On the other hand, it is important for applicability that the platform groups are easy to generate. The process in this case is as follows. First, a set of generators is considered, and then random relators are generated. It should be noted that if the number of generators is *big*, it is quite likely that small cancellation conditions hold for the random relators. After that, the selected relators are symmetrized. Following this, check for which relators the cancellation condition $C'(1/6)$ holds. The ease of generating such groups is a big advantage for small cancellation groups in this context.

In the previously mentioned paper [**27**] another variation of the HKS scheme based in a shortlex order is described (see 4.4 in [**27**]). This method makes the reduction step easier for the participants. The authors state that again the scheme with this variation that $C'(1/6)$ is a good platform (for similar reasons to those above). In addition, they give concrete parameters to generate the groups; they propose groups with forty generators and four relators of length nine. They run several tests (using in particular Staecker implementation [**199**]) to check that the Dehn's algorithm performs successfully the pertinent reductions, and moreover check that every step in the shortlex variation of the HKS algorithm can be undertaken efficiently.

Concerning security, the main problem of this method is that an eavesdropper acquires knowledge on the set of relators corresponding to some participant(s), and this is easier as the number of of times a secret is shared increases. In order to combat this weakness, the authors design another modification of the protocol where the relators that are given to the participants are mixed with new relators, and then almost completely renewed after some updates. This diminishes significantly the danger of interception. This method can be combined with protocols defined in [**105**] in order to keep a secret secure after many updates. The combination of these two methods assure the security of the scheme.

9.4.2. Public-key exchange. In their paper [**197**], Shpilrain-Zapata propose small cancellation groups for a protocol of public-key exchange. After a thorough and sharp analysis of the viability of protocols based on the conjugacy and conjugacy search problems in cryptography, they raise the idea of basing cryptographic schemes in families of groups, rather than in groups as individual entities.

More concretely, these authors define what they call a *commuting action key exchange* (CAKE), which turns out to be a generalization of the classical Diffie-Hellman scheme. The feasibility of the protocol is based on the efficient solvability of the word problem for the groups which are used as platform, while its security relies in the difficulty of solving the conjugacy search problem; in particular, in groups for which the conjugacy problem has great complexity.

As already mentioned, the authors propose small cancellation groups as a platform for this protocol. However, unlike in the schemes that were reviewed in the previous section, finitely presented groups for which the condition $C'(1/6)$ holds are not useful in this context; the reason for this is that these groups are hyperbolic, they possess efficient solutions of the conjugacy problem, as we saw in the corresponding chapter.

However, groups for which the conditions $C(4)$ and $T(4)$ hold turn out to be appropriate in this context. First, as seen in the previous section, the word problem is efficiently solvable for them. On the other hand, although the variation of the Dehn algorithm that was discussed above always solves the conjugacy problem, a polynomial solution of it (and of the conjugacy search problem) is not known for these groups. As these groups can be easily generated in randomly (see previous section), the family of small cancellation groups for which conditions $T(4)$ and $C(4)$ hold is a good platform for the CAKE protocol. A concrete example of group in these conditions is shown in the paper:

$$\langle x, y, z \mid x^2yz^2y^{-1}, y^2zx^2z^{-1}\rangle.$$

The authors also offer a diffusion method, i.e., a way of *disguising* the group elements that are transmitted, in order to make it difficult for an eavesdropper to identify these elements, improving the security of the protocol. The idea is, given a presentation of a group by generators and relations to introduce new generators and perform Tietze transformations in order to obtain a group whose relators have length at most 3.

The reversal of the process, which could be a way in which an eavesdropper can recover the original relators out of the new, is notoriously difficult, especially if the Tietze transformations are not known and the original group is not public.

CHAPTER 10

Polycyclic Groups

10.1. Introduction

Polycyclic groups are natural generalizations of cyclic groups, as they are defined inductively out of cyclic groups via extensions. Although it is an ample family of groups, many properties of them are well known. Such properties include finite presentation, solvable word, conjugacy, and isomorphism decision problems and exponential growth rate in certain subclasses, namely as non-virtually nilpotent polycyclic groups.

Eick and Kahrobaei [**46**] further investigated the algorithmic properties of polycyclic groups. In particular, they explored how the time complexity of the word and conjugacy problems varied with respect to the groups' Hirsch length (see definition below). Their experiments showed that while the time complexity of the conjugacy problem grew exponentially with Hirsch length, the word problem remained efficiently solvable. These results suggested the suitability of polycyclic groups for use in cryptography, and stimulated research into cryptosystems based on these groups and their underlying algorithmic problems.

Because of these favorable algorithmic properties many cryptosystems have been created which employ polycyclic groups as a proposed platform. These include key exchanges such as noncommutative ElGamal and secret sharing via the word problem. In response, heuristic and deterministic methods of cryptanalysis have been developed, including the length-based and linear decomposition attacks. Despite these efforts, there are classes of infinite polycyclic groups that remain suitable for cryptography.

The analysis of algorithms for search and decision problems in polycyclic groups has also been developed. In addition to results for the aforementioned problems we present those concerning polycyclic representations, group homomorphisms, and orbit decidability. Though much progress has been made, many algorithmic and complexity problems remain unsolved; thus polycyclic groups remain a hot topic of research from this point of view.

In this chapter we follow the exposition of [**94**].

10.2. Mathematical background

The nature of polycyclic groups enables them to be represented in several ways. These approaches give rise to complementary algorithms for solving search and decisions problems, with varying degrees of computational complexity. Due to this flexibility, we begin our study of the algorithmic problems in polycyclic groups by examining these representations.

10.2.1. Representations of polycyclic groups.

10.2.1.1. *Polycyclic sequences and Hirsch length.* We recall first some definitions:

DEFINITION 10.1. *A group G is said to be* polycyclic *if it has a subnormal series $G = G_1 \triangleright \cdots \triangleright G_{n+1} = \{1\}$ such that the quotient groups G_i/G_{i+1} are cyclic. This series is called a* polycyclic series. *The* Hirsch length *of a polycyclic group G is the number of infinite factors in a polycyclic series.*

Though a polycyclic group can have different polycyclic series, as a consequence of the Schreier refinement theorem, its Hirsch length is independent of the choice of series. Alt is the notation for the alternating group.

EXAMPLE 10.2 ([**172**]). *Let $G = Alt(4) = \langle (1,3)(2,4),(1,2)(3,4),(1,2,3)\rangle$. Then $V = \langle (1,3)(2,4),(1,2)(3,4)\rangle \triangleleft G$ and $H = \langle (1,3)(2,4)\rangle \triangleleft V$.
Note that $G/V = \langle (1,2,3)\rangle$, $V/H = \langle (1,2)(3,4)\rangle$ and $H \cong \mathbb{Z}_2$. Hence,*

$$G \triangleright V \triangleright H \triangleright 1$$

is a polycyclic series for G.

Consider a polycyclic group G with polycyclic series $G = G_1 \triangleright \cdots \triangleright G_{n+1} = \{1\}$.

DEFINITION 10.2.1. Given that each quotient G_i/G_{i+1} is cyclic, there exists $x_i \in G$ such that $\langle x_i G_{i+1} = G_i/G_{i+1}\rangle$ for $1 \le i \le n$. The sequence of elements $X = [x_1, \ldots, x_n]$ is a *polycyclic sequence* for G.

DEFINITION 10.3. *Let X be a polycyclic sequence for G. The* sequence of relative orders *is $R(X) := [r_1, \ldots, r_n]$ where $r_i := |G_i : G_{i+1}| \in \mathbb{N} \cup \{\infty\}$.*

EXAMPLE 10.4. *Let $G = Alt(4)$. A polycyclic sequence for G is*

$$X = [(1,2,3),(1,2)(3,4),(1,3)(2,4)]$$

and the corresponding sequence of relative orders is $R(X) = [3,2,2]$.

DEFINITION 10.5. *Given a polycyclic group G with polycyclic sequence $X = [x_1, \ldots, x_n]$ and $g \in G$, the expression $g = x_1^{e_1} \cdots x_n^{e_n}$ with $e_i \in \mathbb{Z}$ for $1 \le i \le n$ is the* normal form *of $g \in G$.*

EXAMPLE 10.6. *Let $G = Alt(4)$ with polycyclic series*

$$G \triangleright \langle (1,3)(2,4),(1,2)(3,4)\rangle \triangleright \langle (1,3)(2,4)\rangle \triangleright 1.$$

*Then we may define the polycyclic sequence $X = [x_1, x_2, x_3]$, where $x_1 = (1,2,3)$, $x_2 = (1,2)(3,4)$, and $x_3 = (1,3)(2,4)$. Let $V = \langle x_2, x_3\rangle$ and $K = \langle x_3\rangle$. Given the element $g = (1,2,4) \in G$, we would like to use the polycyclic series to write the normal form of g.
Note that $gV = x_1^2 V$, so that $x_1^{-2} g = (1,4)(2,3) \in V$. In addition, $v := (1,4)(2,3)$ satisfies $vK = x_2 K$, which gives $x_2^{-1} v = (1,3)(2,4) \in K$. Hence, $x_2^{-1} v = x_3$ and we have $v = x_2 x_3$. Putting this information together we have $x_1^{-2} g = x_2 x_3$, which gives $g = x_1^2 x_2 x_3$.*

10.2.1.2. *Polycyclic presentations.* Every polycyclic group can be described by a polycyclic presentation:

$$\begin{array}{lll}\langle g_1,\dots,g_n \mid & g_j^{g_i} = u_{ij} & \text{for } 1\le i<j\le n,\\ & g_j^{g_i^{-1}} = v_{ij} & \text{for } 1\le i<j\le n,\\ & g_i^{r_i} = w_{ii} & \text{for } i\in I\rangle,\end{array}$$

where u_{ij}, v_{ij}, w_{ii} are words in the generators $g_{i+1},\dots,g_n$ and I is the set of indices $i \in \{1,\dots,n\}$ such that $r_i = [G_i : G_{i+1}]$ is finite.

DEFINITION 10.7. *A polycyclic presentation in which every element has a unique normal form is called a* consistent *polycyclic presentation.*

Although not every polycyclic presentation is consistent, there is an effective algorithm to convert an incosistent presentation to a consistent presentation. Unlike general finite presentations, a polycyclic presentation enables the word problem to be solved using an algorithm called *collection* which was introduced by P. Hall (1934), and that we describe below.

DEFINITION 10.8. *Let G be a polycyclic group given by a consistent polycyclic presentation $\langle g_1,\dots,g_n|R\rangle$, where R is the set of relators as defined above. Let g be a word in the generators $g_1,\dots,g_n$ written as $g = g_{i_1}^{e_1}\cdots g_{i_r}^{e_r}$ with $e_j \in \mathbb{Z}$, $i_j \neq i_{j+1}$ for $1 \le j \le r-1$ and $a_j \neq 0$ for $1 \le j \le r$.*

- *A word is* collected *if $g = g_{i_1}^{e_1}\cdots g_{i_r}^{e_r}$ with $i_1 < i_2 < \cdots < i_r$ and $e_j \in \{1,\dots,r_j-1\}$ where r_j is a finite element in the sequence of relative orders; otherwise, g is* uncollected.
- *A word h in the generators $g_1,\dots,g_n$ is a* minimal non-normal subword *of the word g is h is a subword of g and it has one of the following forms:*
  - $h = g_{i_j}^{e_j}\cdot g_{i_{j+1}}$ *for* $i_j > i_{j+1}$,
  - $h = g_{i_j}^{e_j}\cdot g_{i_{j+1}}^{-1}$ *for* $i_j > i_{j+1}$,
  - $h = g_{i_j}^{e_j}$ *for* $r_{i_j} \neq \infty$ *and* $e_j \notin \{1,\cdots,r_{i_j}-1\}$.

From this definition it is clear that a word is collected if and only if it does not contain a minimal non-normal subword.

The collection algorithm works by iteratively applying the power and conjugacy relations of the presentation until the normal form for a word is obtained. The nature of the power and conjugacy relations ensures that the collection algorithm terminates (see [**110**]). The following example illustrates the algorithm.

EXAMPLE 10.9. *Let $G = S_4$ be given by the polycyclic presentation*

$$\langle g_1,g_2,g_3,g_4 | g_1^2 = g_2^3 = g_3^2 = g_4^2, g_2^{g_1} = g_2^2,$$
$$g_3^{g_1} = g_3, g_4^{g_1} = g_3g_4, g_3^{g_2} = g_3g_4, g_4^{g_2} = g_3, g_3^{g_4} = g_3\rangle,$$

where $g_1 = (1,2), g_2 = (1,2,3), g_3 = (1,2)(3,4)$, and $g_4 = (1,3)(2,4)$.

The word $g = g_3g_2g_1$ is uncollected. From the relations, we have

$$\begin{aligned} g_3g_2g_1 &= g_2g_3g_4g_1\\ &= g_2g_3g_1g_3g_4\\ &= g_2g_1g_3g_3g_4\\ &= g_1g_2^2g_3^2g_4\\ &= g_1g_2^2g_4. \end{aligned}$$

The collection can be performed using various strategies, which include *from the left*, *from the right*, and *to the left*. When collection from the left is performed the minimal non-normal subword closest to the beginning of the word is collected; similarly, when collection is performed from the right the minimal non-normal subword nearest the end of the word is collected. Collection to the left, instead, begins by moving all occurrences of g_1 to the beginning of the word followed by all occurrences of g_2 and so on.

In general the collection algorithm is effective in practical applications, but its computational complexity remains unknown.

10.2.2. Growth rate. Given a finitely generated group G, the growth rate is determined by its *growth function* $\gamma : \mathbb{N} \longrightarrow \mathbb{R}$ defined as $\gamma(n) = \#\{w \in G : l(w) \leq n\}$, where $l(w)$ is the length of w as a word in the generators of G. The key space used in a cryptographic protocol should be one that would make an exhaustive search for the key impossible. As such, a fast growth rate of a group G will provide a suitable key space for the protocol.

The class of polycyclic groups which are not virtually nilpotent is known to have an exponential growth rate (see Wolf [**210**] and Milnor [**154**]), making them good candidates for use as platform groups.

10.3. Algorithmic problems

As mentioned in Chapter 2, there are several algorithmic problems in group theory that are utilized as the basis for various cryptographic protocols. The solvability of these decision and search problems in potential platform groups must be extensively examined in order for the use of the groups to be feasible.

For polycyclic groups the following decision problems are decidable:

(a) word decision problem,
(b) single conjugacy decision problem,
(c) subgroup membership decision problem,
(d) isomorphism decision problem.

Remeslennikov [**183**] and Formanek [**64**] showed that the conjugacy decision problem for polycyclic groups is decidable. To see that the word problem is decidable one can observe that it is a special case of the conjugacy decision problem (where $g = u, v = 1_G$), or by observing that every word has a unique normal form induced by a polycyclic presentation. Segal showed that the isomorphism decision problem for polycyclic groups is solvable in [**191**]. Malcev, in turn, proved in [**152**] that the subgroup membership decision problem is solvable for polycyclic groups.

In polycyclic groups, the multiple conjugacy search problem for n elements reduces to n independent solutions of a single conjugacy search [**46**]. Therefore it is acceptable to speak only of the conjugacy search problem without mentioning arity in this setting.

Eick and Ostheimer [**47**] proposed an algorithm for infinite polycyclic groups to solve the conjugacy search problem using polycyclic presentation. The algorithm uses a variety of ideas: it exploits finite orbit and stabilizer computations, calculations in number fields, and linear methods for polycyclic groups. Although the algorithm's complexity is unknown, it seems to work efficiently in polycyclic groups of small Hirsch length.

In [**46**], Eick and Kahrobaei compared the performance of the Eick-Ostheimer algorithm for the constraint satisfaction problem (CSP) against the collection algorithm for polycyclic groups of the form $G = \mathcal{O}_K \rtimes \mathcal{U}_K$, where $\mathcal{O}_K$ and $\mathcal{U}_K$ are respectively the maximal order and group of units of an algebraic number field K. The experimental evidence in [**46**] suggest that the Eick-Ostheimer algorithm becomes impractical as the Hirsch length increases, while the collection algorithm remains efficient. Currently there are no known algorithms for infinite polycyclic groups of high Hirsch length that solve the CSP efficiently, which suggests these groups could potentially be suitable for use as platform groups.

10.4. Cryptographic protocols

Since the field of noncommutative cryptography began, several cryptosystems have been proposed (see Chapter 4) with each of these systems requiring a non-abelian platform group (or other algebraic structure) suitable for cryptography. The minimum criteria for such a group include that the group operations and the word problem must be efficient. As the various protocols are based on different group-theoretical problems, additional criteria will be required depending on the protocol at hand.

As mentioned earlier, polycyclic groups have several properties that make them suitable as cryptographic protocols. While the search for platform groups continues, polycyclic groups are being considered for the following noncommutative protocols:

(a) the Anshel-Anshel-Goldfeld key exchange protocol,
(b) Ko-Lee key exchange protocol,
(c) noncommutative ElGamal key exchange,
(d) noncommutative digital signature (Kahrobaei and Koupparis),
(e) a key exchange using the subgroup membership search problem (Shpilrain and Zapata),
(f) secret sharing schemes based on the word problem (Habeeb, Kahrobaei. and Shpilrain).

The security of these cryptographic protocols rely on various group-theoretic problems as several of the names indicate (see Chapter 4 for details). For convenience, we list them here:

- AAG: simultaneous conjugacy search problem and the subgroup membership search problem,
- Ko-Lee and noncommutative digital signature due to Kahrobaei and Koupparis: conjugacy search problem within subgroups A, B,
- noncommutative ElGamal key exchange: conjugacy search problem in G,
- noncommutative ElGamal where the conjugacy search problem is efficiently solvable in G: power conjugacy search problem in G.

While the secret sharing scheme proposed by Habeeb, Kahrobaei, and Shpilrain relies on an efficiently solvable word problem for feasibility of the scheme, the security of this scheme is contingent upon the relators R_J being kept secret.

As noted in Chapter 5 numerous attacks against group-based cryptosystems have been implemented. Here we will provide an analysis of these attacks on polycyclic groups, beginning with the length-based attack (LBA). The authors of [**67**] analyzed the LBA against AAG over polycyclic groups, determining that the success rate of the LBA decreased as the Hirsch length of the platform group increased.

The authors implemented LBA with Memory 2 in their study. Although the groups were resistant to LBA, it was later shown (see below) that these groups were susceptible to a field based attack.

The LBA functions, like the platform group G, are a free group when attempting to recover the secret from the public information in the AAG protocol. The success of the LBA in groups which were not free motivated Miasnikov and Ushakov in [**164**] to investigate the asymptotic properties of the given platform groups. Their investigation determined that the LBA is successful for groups where a random choice of elements will likely generate a free subgroup of G.

This led to a new form of attack, deemed the *quotient attack*, for the AAG key exchange protocol, as well as others, that use some variation of the membership or conjugacy search problems. The algorithm solves the search problems in a quotient group G/N. If G/N possesses the exponentially-generic free basis property, the solution in the quotient will yield one in the original group (see Chapter 5). Since polycyclic groups do not possess the free basis property nor any free subgroups, this attack is not applicable to polycyclic groups.

The linear decomposition attack (see Chapter 5), introduced by Miasnikov and Roman'kov, provides a general framework for the cryptanalysis of many group-theoretic analogues of the Diffie-Hellman key exchange. For a protocol to be susceptible to the attack, its platform group must admit a linear representation and the algorithmic security assumption of the protocol must be equivalent to commutative linear transformations. In order for the linear decomposition attack to be implemented, the platform group must be given by its linear representation V or by by a presentation in conjunction with a faithful embedding into $GL(V)$. In addition, the group must be embedded in a linear space of sufficiently small dimension for the attack to be viable. While the dimension of the smallest linear embedding of finite groups and some classes of infinite groups (for example torsion-free nilpotent and polycyclic-by-finite) are known, the authors acknowledge that no such bounds are known for other linear groups, including general polycyclic groups and metabelian groups.

In [**137**], Kotov and Ushakov investigated the security of the AAG key exchange protocol using the polycyclic groups suggested by Eick and Kahrobaei in [**46**]. Recall that these were groups of the form $G_F = \mathcal{O}_F \rtimes U_F$, where $\mathcal{O}_F$ is the maximal order and U_F is the unit group generated by an irreducible polynomial in the algebraic number field F. Here, U_F acts on $\mathcal{O}_F$ by right multiplication. The authors constructed the field-based attack, which is able to recover the shared key regardless of the group's Hirsch length in polynomial time. The key is recovered by solving a linear system of conjugacy equations over the field F. In order for the attack to be applied directly, the group G_F must be specified as a semidirect product and F is given in matrix form. When G_F is given by a polycyclic presentation rather than matrix form, the authors construct a linear representation utilizing the polycyclic presentation. Then this linear representation is used to recover the shared key.

Although the field-based attack is successful in these particular groups, there are other classes of polycyclic groups that can be resistant to such an attack. Since the protocol relies on the linear representation of the groups, polycyclic presentations in which the matrix representations are not readily computable could be good candidates for the AAG protocol.

The linear centralizer attack (see Chapter 5) against AAG due to Tsaban [**205**] uses a faithful linear representation of a braid group $\mathcal{B}_n$. Although the algorithm is polynomial time with respect to n, it is impractical for small values of n. The linear centralizer attack is applicable to groups whose faithful linear representations are known and sufficiently small. Hence, as these properties for polycyclic groups are still unknown, polycyclic groups can still be resistant to these attacks.

CHAPTER 11

Graph Groups

11.1. Introduction

In this chapter we will study graph groups, which are also referred to as partially commutative groups or right-angled Artin groups in the literature. These objects, introduced by Hauschild and Rautenberg [**100**], have been widely studied since then, as they are situated in a crossroad between group theory, geometry, graph theory, and complexity theory. Although they are easily defined (see below) they possess a very rich subgroup structure that in particular allows the construction of cryptographic platforms based on them. Moreover, as we will see below, their decision problems have been thoroughly studied, and solved in many cases, in part because of the existence of manageable normal forms for the elements of these groups.

It is important to remark that every graph group is defined in a unique way by means of a graph, which is finite if the group is finitely generated. Then every algorithmic problem regarding graphs can be translated to the language of graph groups, and in this way the algorithmic complexities of these objects are related. Note that graph-theoretic problems have been central in the development of complexity theory, so this bridge permits one to deal with these problems using tools of group theory. Because of this there is an important line of research whose goal is to express statements and results of graph theory in terms of graph groups, without explicit reference to concrete systems of generators if possible (see [**63**] for a compilation of these achievements).

This chapter is structured as follows. First we recall the basic definitions about graph groups, including the background that will be necessary in the remainder of the chapter. Then we review results that have been achieved concerning decision problems in these groups, describe cryptosystems that have been developed using them as platforms, and we conclude with a list of open problems.

The main source for this chapter has been the paper [**61**].

11.2. Mathematical background

In this section we introduce the graph groups. We start by defining them in terms of the graph which is the motivation of the name, though the reader can check that an abstract definition of the groups is easily deduced. We will restrict ourselves to the finite case, which will be sufficient for our purposes.

DEFINITION 11.1. *Let* Γ *be a finite graph,* $V(\Gamma)$ *its set of vertices and* $E(\Gamma)$ *its set of edges. The* graph group *defined by the graph* Γ *is the group:*

$$A(\Gamma) = \{V(\Gamma) \mid [v_i, v_j] \textit{ if } (v_i, v_j) \in E(\Gamma)\}.$$

The generators associated with the vertices of Γ are called *special* or *vertex generators* of $A(\Gamma)$, and are usually denoted by the same letters as the vertices. In the definition, $[v_i, v_j]$ stands for the commutator of v_i and v_j. Two finite graphs are isomorphic if and only if the associated graph groups are isomorphic [**45**].

Here we provide some examples.

- If Γ is the empty graph of n vertices, then $A(\Gamma)$ is the free group in n generators.
- If Γ is a clique of n-vertices, then $A(\Gamma)$ is the free abelian group in n generators.
- If Γ is a chain with three vertices, then $A(\Gamma)$ fits in a central extension $\mathbb{Z} \to A(\Gamma) \to \mathbb{Z} * \mathbb{Z}$.
- If Γ is given by two disconnected edges, then $A(\Gamma) = \mathbb{Z}^2 * \mathbb{Z}^2$.

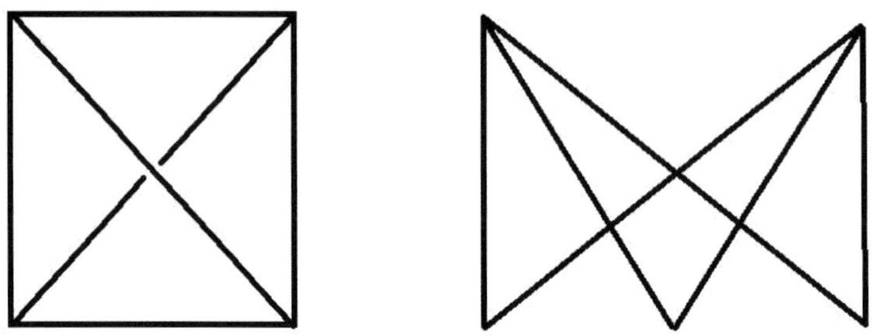

FIGURE 11.1. The graphs K_4 and $K_{2,3}$, which correspond to the groups $\mathbb{Z}^4$ and $F_2 \times F_3$, respectively.

From the point of view of the next section, it is very important that there is always at least a normal form for the elements of a graph group $A(\Gamma)$; let us recall how it can be constructed. First, choose an order of the vertices of Γ. Let a be an element of $A(\Gamma)$, and consider the set M of words of minimal length that represent a. By a result of Green [**81**], the vertices that can occur as the initial letter of a define a free abelian subgroup of $A(\Gamma)$. Let $T \subseteq A$ the graph generated by these vertices, which is necessarily complete. Now "moving to the left" in a we consider the elements of T that can appear as the initial letter of a to obtain that $a = w_0 a_1$, with $w_0 \in T$ of maximal length. Iterating the process for a_1, we obtain that $a = w_0 w_1 \dots w_j$ in an unique way, up to commutation inside a certain T_i. If we impose additionally that the letters of each w_i respect the initial order of the generators, we obtain that the form $w_0 w_1 \dots w_j$ is unique. For more information we refer the reader to [**104**].

EXAMPLE 11.2. *In this example, consider the lexicographic order in the generators, and consider the element* $a = x_3 x_1 x_2 x_1^{-1}$. *We are going to describe the normal form of this element for the four groups defined in the previous example.*

- *If* $A(\Gamma) = F_3$, *no letters in the previous expression of* a *commute, so the normal form is precisely* $a = x_3 x_1 x_2 x_1^{-1}$.
- *If* $A(\Gamma) = \mathbb{Z}^3$, *all the letters commute, and hence the normal form is* $x_2 x_3$.
- *If* $A(\Gamma) = \langle x_1, x_2, x_3 \mid [x_1, x_2], [x_2, x_3] \rangle$, *note that* x_2 *and* x_3 *can appear as the initial letter of* a, *but* x_1 *cannot as it does not commute with* x_3. *Hence, the normal form is again* $x_2 x_3$.

- *If* $A(\Gamma) = \langle\langle x_1, x_2, x_3, x_4 \mid [x_1, x_3], [x_2, x_4]\rangle$, *then* x_3 *and* x_1 *can occur as the initial letters of* a, *but neither* x_2 *nor* x_1^{-1}. *Then, the normal form is* $x_1x_3x_2x_1^{-1}$.

Other group-theoretic, geometric, and topological features of graph groups are well-studied and described, as for example:

- Centralizers
- Automorphisms
- Quasi-isometries
- Cohomology over fields and over the integers
- Actions over cubical complexes
- Models for Eilenberg-MacLane spaces
- Structure of their associated Coxeter groups
- Families of subgroups.

We refer the interested reader to the excellent surveys [**29**] and [**133**], and references therein.

11.3. Algorithmic problems

In this section we describe the different attempts that have been undertaken in order to solve the main algorithmic problems for graph groups.

11.3.1. Word problem. The first solution of the word problem for graph groups used normal forms and corresponded to Servatius [**193**] in a paper devoted to the study of the automorphisms of a graph group. He also solved the conjugacy problem for these groups (see below), but he did not give any hint about the complexity of the problem.

Some years later, Wrathall proved in [**211**] that the word problem for graph groups is solvable in linear time. More concretely, she proves that given a presentation of the graph group, the presentation of the associated monoid is preperfect as a Thue system, i.e., a set of pair of words subject to some closure properties. As a consequence, there exists an algorithm that, for every word in the group, produces a normal form as a projection in linear time.

These methods for finding efficient solutions of decision problems in monoids are generalized in [**147**], where Liu-Wrathall-Zeger solve and give the complexity of the generalized word problem for graph groups:

THEOREM 11.3 ([**147**], Corollary 4.3). *Given to elements* x *and* y *in a graph group, it can be decided if there exist* n *and* m *such* $x^n = y^m$, *and the complexity of the algorithm is linear in the length of* xy.

Hence, it should be considered that the word problem in graph groups can be solved very efficiently.

11.3.2. Conjugacy problem. The first results concerning the conjugacy problem for graph groups run parallel to those of the word problem that we have just discussed. In the aforementioned paper of Servatius [**193**] it is also established that the conjugacy problem is solvable for these groups, while Lee-Wrathall-Zeger [**147**], making use of the corresponding result for monoids, prove that is in fact solvable in linear time. The work of van der Wyck should also be mentioned here, as this author shows that these groups are biautomatic [**206**].

A more recent contribution regarding this subject is the paper of Crisp-Godelle-Wiest [**37**]. By extending the Viennot and pyramidal piling to the context of graph groups a new normal form is obtained. They provide a new proof of the solvability of the conjugacy problem in linear time, resulting in:

THEOREM 11.4 ([**37**], Theorem 2.22). *The conjugacy problem in a graph group is linear time in the sum of the lengths of the two elements involved.*

The proof essentially consists of producing the pilings for the two elements and then factoring them in nonsplit factors, producing the pyramidal pilings of the factors. Then a corresponding normal formal is constructed out of them.

Using these methods, the authors are also able to prove the linearity of the conjugacy problem for a far more general class of groups that appear as subgroups of graph groups; namely, fundamental groups of cubical complexes as presented in Haglund and Wise [**98**]. This class contains, in particular, the graph braid groups [**55**] and many hyperbolic groups. On the other hand, Bridson establishes in [**19**] that the conjugacy problem is not always solvable for finitely presented subgroups of graph groups. His construction is based on the existence of a group of F_3-type with unsolvable word problem, which is a torsion-free quotient of a hyperbolic virtually special torsion-free group.

11.3.3. The isomorphism problem. Recall that given two different graphs, the graph isomorphism problem asks whether they are isomorphic or not. The problem is known to be subexponential, and Babai has announced a proof that it is quasi-linear [**5**], which is not published yet but generally believed to be correct after some corrections from Helfgott (known as the Babai-Helfgott solution). As explained above, two graph groups are isomorphic if and only if their corresponding graphs are, so these statements apply verbatim for the isomorphism problem for graph groups.

If we consider subgroups of graph groups, it has been proved by Bridson that the isomorphism problem for them is not solvable in general, without considering the hypothesis of being finitely generated [**19**]. In his proof, he constructs a graph group A and a finite-index subgroup P_0 such that, by a criterion of Bridson and Miller [**135**] based on the Rips complex, the isomorphism problem is not solvable for subgroups of $P_0 \times P_0 \times F$, where F is a certain free group. Hence, the problem is also unsolvable for the graph group $A \times A \times F$.

Using similar methods, Bridson also proves that the subgroup isomorphism problem is unsolvable for graph groups. For the interested reader, we recommend the recent work of Casals-Ruiz [**25**].

11.3.4. The group homomorphism problem. As in the previous case, this is a version of the analogous problem for graphs: given two graphs Γ and Γ', does there exist a graph homomorphism $\Gamma \to \Gamma'$? However, and unlike above, to define the corresponding problem in graph groups requires two restrictions:

- There should exist sets of special generators $S \subset A(\Gamma)$ and $S' \subset A(\Gamma')$ such that for the homomorphism $f : A(\Gamma) \to A(\Gamma')$, $f(S) \subseteq f(S')$.
- If x and y belong to S and commute, their images by f should also commute.

Under these two specifications, the group homomorphism problem for graphs and group graphs are equivalent. According to [**144**], the problem is equivalent to

k-colorability and is NP-complete as long as three elements of S' generate a free abelian group inside $A(\Gamma')$.

11.3.5. The membership problem. Following the recent research concerning subgroups of graph groups, there has been certain interest in the membership problem for this class of groups. Recall that given an inclusion $H < G$ of subgroups, the subgroup H is said to be *undistorted* if the embedding is quasi-isometric with respect to the word metric. Moreover, if $|h|_T$ is the length of h in T (respectively in S), given a system of generators T of H and a system of generators S of G such that $T \subseteq S$, the *distortion function* $D : \mathbb{N} \to \mathbb{N}$ takes every n to $\max\{|h|_T \text{ such that } |h|_S \leq (n)\}$.

It was shown in [**61**] that for every finitely generated undistorted subgroup of a graph group, the membership problem is solvable in at most exponential time; in fact, this result is true whenever the word problem of the big group is solvable in at most exponential time. Moreover, it is proved that there exists a graph group G and an exponentially distorted group $H < G$, which is a surface group, such that the membership problem is solvable in exponential time. It is in fact conjectured that the complexity could be linear.

On the other hand, Bridson has proved, using similar examples to those discussed in the previous sections, that there exists subgroups of graph groups for which the membership problem is unsolvable.

11.3.6. Geodesic problems. In [**160**, Section 18.3] it is shown that the geodesic length problem, the geodesic problem, and the bounded geodesic length problem are polynomially equivalent to each other. Computing in logspace and taking advantage of previous computations performed in Coxeter groups, a normal form for graph groups is obtained in quadratic time, which is in particular a geodesic word. Hence, the three problems are solvable in polynomial time for these groups.

11.4. Cryptographic protocols

In this section we will describe some cryptographic protocols that have been described using graph groups as a platform, and take advantage of the interesting algorithmic features of these groups that were described in the previous section.

Historically, the first cryptosystem that used graph groups was described by Wagner-Magyarik in [**207**]. Their proposal was based on a group $\langle S \mid R \rangle$ where the word choice problem was hard, but it became easy when new relators were added to S. As they mentioned, graph groups fit well in this context. Later, it was shown that their system was vulnerable to a reaction attack. This problem was corrected by Levy-Perrett [**145**], who switched the platform from graph groups to partially commutative monoids.

There have been several proposals of secret sharing schemes that make use of graph groups. In [**60**], the easy solvability of the word problem for these groups allows the use of them as platforms for the secret sharing threshold schemes defined by Habeeb-Kahrobaei-Shpilrain and already described in Chapter 4. Recall that the security of this scheme strongly depends on maintaining secret the relations of the groups involved.

In [**61**], a straightforward secret sharing scheme that depends on the solvability of the group decomposition problem for graph groups is proposed. Let us briefly sketch it. Given n participants, a dealer gives shares of the generators and relations of a graph group G_i to each participant P_i. Now each participant P_i gets a number m_i which is 1 if the group G_i decomposes as a direct product, and 0 otherwise. On the other hand, there is a monic polynomial f of degree n, which is publicly unknown, such that $f(i) = m_i$ for every $1 \leq i \leq m_i$. Now the knowledge of all the $m_i's$ allows us to obtain the secret key, which is $f(0)$. It should be remarked that this scheme is considered an elementary idea and can be perfected in order to improve the security, using the classical ideas of Shamir.

Authentication schemes based on graph groups have also been described. We present the proposals of the aforementioned article [**61**], based on work of Grigoriev-Shpilrain [**91**] for the case of graphs.

In the first scheme, two finite presentations $\langle S_i \mid R_i \rangle$ of respective graph groups G_i, $i = 1, 2$, are the public key of Alice. Her private key is a homomorphism $\alpha : G_1 \to G_2$ such that $\alpha(S_1) \subseteq S_2$. Now Alice selects another graph group G given by the presentation $\langle S \mid R \rangle$ and a homomorphism $\beta : G \to G_1$ such that $\beta(S) \subseteq S_1$. She sends the group G to Bob. Then Bob sends a random bit c to Alice; if $c = 0$, Alice sends β to Bob, and Bob should check that $\beta(G) = G_1$ and $\beta(S) \subseteq S_1$. If $c = 1$, Alice sends the composite $\alpha\beta$ to Bob, and he checks if $\alpha\beta(G) = G_2$ and $\alpha\beta(S) \subseteq G_2$.

In the second version, two isomorphic subgroups G_1 and G_2 of a graph group G are the public key of Alice, while the private key is an isomorphism α between them. Now Alice selects another group isomorphic to G_1 and sends it to Bob, keeping an isomorphism $\beta : G \to G_1$ as her private key. Bob sends a random bit c to Alice. If $c = 0$, Alice sends β to Bob, and he checks that it is an isomorphism between G and G_1. On the other hand, if $c = 1$, Alice sends $\alpha\beta$ to Bob, and he checks that this composite is an isomorphism between G and G_2.

The security of these two authentication schemes depends on the unsolvability (in general) of the group isomorphism and group homomorphism problems, discussed in the previous section.

11.5. Interesting problems

In this section we give a list of problems that concern decision problems in graph groups and applications to cryptography. A number of them are, with nuances, translations to this context of problems in graph theory, using the bridge that is provided by the graph of the group as explained above. Many of the problems admit multiple variations. Our source here is the last section of [**61**].

- Given a graph group G and a system S of special generators, find the subsets of S that generate free abelian subgroups of G of a fixed rank k.
- The analogous problem to the one above by writing *free* instead of *free abelian.*
- Given graph groups G_1 and G_2 with respective systems of vertex generators S_1 and S_2, check if there exists an inclusion $G_1 \hookrightarrow G_2$ such that $i(S_1) \subseteq S_2$.
- Given graph groups G_1 and G_2 with respective systems of vertex generators S_1 and S_2, decide if G_1 is a subgroup of the group obtained by adding in succession new vertex generators to G_2 which commute with exactly two generators of the previous step.

- Consider a graph group $G = \langle S \mid R \rangle$, and a partition $R_1, \ldots, R_m$ of the set of commutator relations. Then the *chromatic number* of the graph group can be defined as the minimum cardinal of a partition of R such that no R_i contains two commutators that involve the same vertex. There is a substantive literature concerning coloring of graphs that can be translated near verbatim to graph groups. For example, the problem of finding the chromatic number of a graph group is NP-complete, as a consequence of the analogue problem for graphs. It would be interesting to know if the Polly Cracker scheme defined in [**135**] can be reformulated in terms of graph groups. From a different point of view, coloring of graph groups is used in [**129**] in a thorough study of the embedability between these groups.
- Consider a graph group $G = \langle S \mid R \rangle$. The *$k$-vertex cover problem* consists, in this context, of finding a subset $S' \subseteq S$ with $|S| = k$ such that every vertex generator of S commutes with at least one element of S'.
- Consider again a graph group $A(\Gamma) = \langle S \mid R \rangle$, and a collection $\{S_i\}$ of distinct subgroups of S such that their union is S, and such that the subgraph of Γ associated to S_i does not contain a cycle (note that this condition can be formulated in purely group-theoretical terms). Then the problem is to find the minimal cardinal of such a collection, called the *arboricity* of $A(\Gamma)$.
- In general, which other problems in graph theory can be translated to the language of graph groups in a way that they yield interesting complexity results over these groups?
- Is it possible to translate to graph groups the secret sharing scheme defined in [**138**] by means of graph coloring?

CHAPTER 12

Arithmetic Groups

12.1. Introduction

In this chapter we describe the proposal of [**118**] that pioneered the use of arithmetic groups as platforms for cryptographic protocols. These groups can be described as the groups of integer points of linear algebraic groups defined over the rational numbers. Although related ideas go back to the work of Hermite and Minkowski in quadratic forms, their systematic study was developed in the sixties and seventies in the work of Borel, Weil, and Margulis, among others. Aside from being a central object in (algebraic) number theory, they play an important role in different subjects such as Riemannian geometry, operator theory, Lie theory or the Langlands program. However, and probably because combinatorial descriptions of these groups are often not easily available, they have not played a significant role in cryptography thus far.

We will establish that some instances of arithmetic groups, for example the special linear groups over the integers, offer a good platform for a symmetric-key exchange protocol. This protocol was first described by Chatterji-Kahrobaei-Lu in [**30**], and is based on the existence of subgroups which are exponentially distorted in the big group and have an efficient solution of the geodesic length problem. It will be shown that these properties hold for some unipotent groups of arithmetic groups (see definition below). Moreover, it is important to point out that the same scheme it is also valid for some arithmetic-like groups, see Theorem 12.6.

The contents of this chapter are organized as follows. We start with the mathematical background necessary to understand the arithmetic groups involved in the protocol. As always in this book, the section is intended to be comprehensive, concise and as self-contained as possible. Then, we will develop the behavior of these groups with respect to some algorithmic problems, and we finish by discussing the cryptographic protocol and its security.

This chapter is mainly based in the results of [**118**], whom we refer the reader for details and proofs.

12.2. Mathematical background

We start by recalling the definition of linear algebraic group, which naturally leads to the concept of arithmetic group. A thorough reference for this topic is [**18**].

DEFINITION 12.1. *Let k be a field and G a subgroup of $GL(n,k)$ for some $n>0$. The group G is a* linear algebraic group *if it can be defined by means of polynomial equations with coefficients in k.*

In the following, we will always take $k=\mathbb{Q}$, the rational numbers.

Recall that two subgroups G and G' of a group H are said *commensurable* in H if H/G and H/G' are finite. Now we can define arithmetic groups.

DEFINITION 12.2. *A linear algebraic group $G < GL(n, \mathbb{Q})$ is said to be* arithmetic *if it is commensurable, with respect to any embedding $G \to GL(n, \mathbb{Q})$, to another linear algebraic group $H < GL(n, \mathbb{Q})$ whose elements are matrices with integer entries.*

Observe that arithmetic groups are always *discrete.*

The next definitions are crucial in this paper, as the protocols are mainly constructed out of (virtually) unipotent subgroups of arithmetic and arithmetic-like groups.

DEFINITION 12.3. *Let G be an arithmetic group. An element $g \in G$ is called* unipotent *if $(g-1)^n = 0$ for some $n > 0$. Equivalently, g can be expressed as an upper triangular matrix where the elements of the main diagonal are all equal to 1.*

An element g is called virtually unipotent *if some power of it is unipotent.*

We will be interested in groups for which all the elements are (virtually) unipotent.

DEFINITION 12.4. *A subgroup of an arithmetic group is* unipotent *(resp.* virtually unipotent*) if all its elements are unipotent (resp. virtually unipotent).*

The definition of unipotence in terms of matrices implies that in general there are many unipotent elements in the arithmetic groups.

In the paper [**118**] the notion of U-element is also introduced:

DEFINITION 12.5. *Let G be an arithmetic group with a given presentation. It will be said that an element $g \in G$ of infinite order is a U-element if, for every $n > 0$ and some constant C, the geodesic length of g^n in terms of the generators of G is less than or equal to $C log(|n| + 1)$.*

Observe that in particular $|n|$ is the geodesic length of the element g^n in the group $\langle g \rangle$ with respect to the generator g. Then, being a U-element is equivalent to generating an infinite, exponentially distorted subgroup of G.

The relation between (virtually) unipotent elements and U-elements is given by the following result, which will be very important in the definition of the protocol. We state here the concrete case that will be useful in this chapter, although the result is more general:

THEOREM 12.6 ([**118**], Theorem 1.1). *Let $n \geq 3$. Then an element of $SL(n, \mathbb{Z})$ or $SL(n, \mathbb{Z}[x_1, \ldots, x_m])$ is a U-element if and only if it is virtually unipotent.*

Hence, these arithmetic groups possess many exponentially distorted subgroups, which is a desirable fact in relation to the protocol that we intend to apply.

Although the protocols that are defined here are likely to be applied in a general class of arithmetic groups, we will mainly deal with the special linear cases, which are especially appropriate for applications.

To finish this section, we will consider a distinguished subgroup $H < SL(n, \mathbb{Z})$, which will be important for computations, and prove a crucial property of its normalizer. We assume $n \geq 3$.

We denote by $H_n < SL(n, \mathbb{Z})$ the subgroup of upper triangular matrices such that $h_{ii} = 1$ if $1 \leq i \leq n$, $h_{ij} = 0$ if $i \neq j$ and $j \neq n$. In other words, only nontrivial elements are on the main diagonal (which are all equal to one) and in the last column. This is clearly an algebraic subgroup of $SL(n, \mathbb{Z})$, as the entries of the matrix elements of H_n are precisely the solutions of the polynomial equations $x_{ii} - 1 = 0$, for $1 \leq i \leq n$, and $x_{ij} = 0$, for $i \neq j$ and $j < n$. Observe that by performing the appropriate transformations on these equations, it can be easily proved that every conjugate of H_n in $SL(n, \mathbb{Z})$ is again an algebraic subgroup.

PROPOSITION 12.7 ([**118**], Proposition 3.4). *The normalizer of H_n has infinite index in $SL(n, \mathbb{Z})$.*

PROOF. A direct computation using the definition of normalizer shows that a matrix $M \in SL(n, \mathbb{Z})$ is an element of $N_{SL(n,\mathbb{Z})} H_n$ if and only if has the shape

$$\begin{pmatrix} B & x \\ 0 & \pm 1 \end{pmatrix},$$

where B belongs to $GL(n-1, \mathbb{Z})$ and $x \in \mathbb{Z}^{n-1}$. This already implies that the normalizer is infinite because, for example, all the lower triangular matrices of $SL(n, \mathbb{Z})$ such that $a_{21} \neq 0$, $a_{ii} = 1$ $(1 \leq i \leq n)$ and $a_{ij} = 0$ in any other case give different classes in $SL(n, \mathbb{Z})/N$. □

12.3. Algorithmic problems

In this section we will consider the algorithmic problems in arithmetic groups that will be useful when applying the protocol defined in 8.4.

12.3.1. A geodesic length problem. Consider a group G given by a presentation $\langle S \mid R \rangle$, and a subgroup $H < G$. In [**118**] a variation of the classical geodesic length problem was proposed and solved in the necessary case. The variation is stated: Is there a positive constant $C > 0$ such that for every $h \in H$ it is possible to construct (in an efficient way) a path in the Cayley graph of G from the trivial element to h, such that the length of this path is bounded above by $C_1 |h|_G$?

Now we describe the solution given to this problem for the case $G = SL(3, \mathbb{Z})$ and $H = H_n$, the upper triangular subgroup defined in the previous section. It is noteworthy that the same result and proof are valid for $SL(n, \mathbb{Z})$, and the case $n = 3$ is presented for the sake of simplicity. In the statement we keep notation from the previous paragraph.

PROPOSITION 12.8 ([**118**], Proposition 3.3). *Let h be an element of $H < SL(3, \mathbb{Z})$, v the vector corresponding to the last column. Then the length of h is uniformly bounded by the number $C log(1 + ||v||)$ for some constant C. Moreover, there is always a path of length $C log(1 + ||v||)$ that can be constructed from the identity to h in polynomial time in $||v||$.*

Sketch proof. The proof is based on the fact that for the eigenvalues λ_1 and λ_2 of the matrix

$$\begin{pmatrix} 1 & 1 \\ 1 & 0 \end{pmatrix},$$

which are the golden ratio and its negative reciprocal, it holds that $\lambda_1 > 1$ and $-1 < \lambda_2 < 0$. Using this, it can be proved that every vector in $\mathbb{Z}^2$ has bounded distance to the set

$$\left\{ \sum_{k=-m}^{-1} \beta_k A^k e_2 + \sum_{k=0}^{m} \beta_k A^k e_1, \text{with } \beta_i \in \{0, 1, 2\}, m \geq 1 \right\}.$$

Here $\{e_1, e_2\}$ is the canonical basis. Now, given v, the value of m can be chosen of the order of $C\log(1 + ||v||)$, and the construction implies that a path connecting the identity and h can be found in polynomial time in $||v||$.

The variant of the geodesic length problem discussed at the beginning of the section is a consequence of the previous result. To see this, observe that, up to a constant, the natural (Riemannian) metric in $SL(3, \mathbb{Z})$ is equivalent to the word metric in $SL(3, \mathbb{Z})$.

12.3.2. Membership problem. Recall that given a subgroup $H < G$, the membership problem asks if given an element $g \in G$, it is possible to decide if h belongs to H. In our case of interest, with $G = SL(n, \mathbb{Z})$ and $H = H_n$, it is clear that solving the linear equations that define the elements of H_n (presented in the previous section) is a problem of polynomial complexity of the size of the entries of h. A similar argument, taking into account that any conjugate of H_n is determined by conjugacy of the equations proves that the membership problem is also solvable in polynomial time for the conjugate subgroups.

12.4. The protocol

The symmetric key exchange for arithmetic groups, and particularly integral special groups, is described in Section 8.4, so we will not repeat it here. We do recall however the properties that must hold for the pair (G, H) in order to serve as a good platform for the scheme:

- H is exponentially distorted in G.
- The membership problem is efficiently solved for H.
- There are many different conjugates of H inside G.
- The variation of the geodesic length problem described in the previous section is solvable in polynomial time.

All of these conditions hold when G is a special linear group over the integers and H is the unipotent group in this setting. Moreover, it can be checked that these conditions also hold for pairs where G is some "arithmetic-like" group such as $SL(n, \mathbb{Z}[x_1, \ldots, x_m])$ and H is a unipotent subgroup. The key to prove this is that the Lubotzky-Moses-Raghunathan theorem also holds for these groups; see [**149**] for a proof. Special groups over distinguished quotients of integer polynomial rings are also good candidates to be platforms for the protocol, since U-elements, as well as exponentially distorted subgroups can be easily identified on them from the U-elements of $SL(n, \mathbb{Z}[x])$.

We conclude this section with some comments on the security of the protocol. The first topic concerns the possibility of a known-plaintext attack (KPA). This attack consists, in our context, of an eavesdropper recovering enough elements whose length (in an unknown generating set of G) is known, which leads to the knowledge of the secret generating set. The following proposition proves that the scheme is secure against this kind of attack.

PROPOSITION 12.9 ([**118**], Proposition 5.1). *Consider any infinite finitely generated subgroup $\Gamma < GL(n, \mathbb{C})$. Consider elements $\{g_1, \ldots, g_m\}$ of Γ and natural numbers $\{k_1, \ldots, k_m\}$ such that the geodesic length of every g_i, with respect to a certain generating set $S \in L$, is k_i. Then there is an infinite number of generating sets S_j of Γ for which this condition holds for the same g_i's and k_i's, and such that the geodesic metrics of S_{j_1} and S_{j_2} are different if $j_1 \neq j_2$.*

Idea of the proof. The proof consists in two parts: in the first it is shown that for every $j \geq 1$ there exists $t_j \in \Gamma$ such that $t_{j_1} \neq t_{j_2}$ if $j_1 \neq j_2$, and that the condition over the $g_i's$ and the $k_i's$ holds for all the generator systems $S_j = S \cup \{t_j\}$. Then, a new infinite sequence of generators systems $\{\bar{S}_i\}$ is defined inductively, with $\bar{S}_1 = S_1$, $\bar{S}_n = \bar{S}_{n-1} \cup \{t_n\}$. By the first part, the first condition of the statement holds, and the second is a consequence of the fact that the geodesic length of t_j is bigger than 1 in $\bar{S}_{j-1}$, and exactly 1 in $\bar{S}_j$, for every j.

On the other hand, it is worth mentioning that given the pair $(SL(n, \mathbb{Z}), H_n)$, as defined above, every conjugate of the group H_n is exponentially distorted in the special linear group. Since all the items mentioned at the beginning of this section hold for such a conjugate, we obtain that for every $g \in SL(n, \mathbb{Z})$, the pair $(SL(n, \mathbb{Z}), gH_ng^{-1})$ is also a good platform for the symmetric key exchange. In general, the choice of an appropriate conjugate of H_n may improve the security of the platform, as it will be harder to identify than H_n.

CHAPTER 13

Engel Groups

13.1. Introduction

In this chapter we will present the current state of Engel group-based cryptography. In particular, we investigate several group-theoretic problems in Engel groups, with a view towards their applications to cryptography (see [**119**]).

The study of Engel groups originates from the famous paper of Burnside in 1901 and it is closely related to the general Burnside problem (in the reminder GBP, for short) [**24**]. The general Burnside problem asks if a finitely generated torsion group is finite. The question whether a finitely generated (n-)Engel group is nilpotent is the analogue of the GBP in the realm of Engel groups. In the following, we refer to this question as the "Engel-Burnside problem". Of course, every group that is locally nilpotent is an Engel group, but the converse needs not be true in general. The first main result on Engel groups is the theorem of Zorn which shows that any finite Engel group is nilpotent.

The contents of this chapter are organized as follows. In Section 13.2 we give basic definitions and properties of Engel groups. We provide some identities satisfied by n-Engel groups, and, in particular, we survey known and open questions about the Engel-Burnside problem. In Section 13.3 we describe a number of cryptosystems that have been built around Engel groups. In particular, we present a public-key cryptosystem based on 2-Engel groups and a digital signature based on 4-Engel groups. Moreover, we survey the learning problem homomorphism which uses as platform group the Burnside groups of exponent 3 (that are, in particular, 2-Engel groups).

The last section is devoted to surveying the current status of Engel groups in cryptography, and we include a list of open problems.

13.2. Mathematical background

Let G be a group and let $x_1, x_2, \ldots$ be elements of G. Recall that the commutator $[[x_1, \ldots, x_{n-1}], x_n]$ of weight $n \geq 1$ was defined in Section 1.8.

If $x = x_1$ and $y = x_2 = \cdots = x_{n+1}$, we use the shorthand notation

$$[x, {}_n y] = [x, y, \overset{n}{\ldots}, y].$$

In the following we recall the definition of Engel elements of a group.

Definition 13.1. *Let G be a group and $g \in G$. We say that g is a* right Engel element *if for any $x \in G$ there exists $n = n(g, x) \geq 1$ such that $[g, {}_n x] = 1$. If n can be chosen independently of x, then g is a* right n-Engel element. *Similarly, g is a* left Engel element *if for any $x \in G$ there exists $n = n(g, x) \geq 1$ such that $[x, {}_n g] = 1$. Again, if the choice of n is independent of x, then g is a* left n-Engel element.

The sets of right and left Engel elements of G are denoted by $R(G)$ and $L(G)$, respectively. Notice that $R(G)$ and $L(G)$ are invariant subsets under automorphisms of G. Hence, for any $g \in R(G)$ (resp. $g \in L(G)$) and any $x \in G$, we have $g^x \in R(G)$ (resp. $g^x \in L(G)$).

The following gives the iteration between right and left Engel elements.

PROPOSITION 13.2 ([**102**]). *Let G be a group. We have*

$$R(G)^{-1} \subseteq L(G) \text{ and } R_n(G)^{-1} \subseteq L_{n+1}(G).$$

Notice also that in every 2-group, elements of order 2 are left Engel, as the following proposition shows (see [**186**] for a proof).

PROPOSITION 13.3. *Let G be group and let $g \in G$ such that $g^2 = 1$. Then for any $x \in G$ and any $n \geq 1$:*

$$[x, {}_n g] = [x, g]^{(-2)^{n-1}}.$$

In particular, every involution (i.e., an element of order 2) in any 2-group is a left Engel element.

In the following we give definitions of Engel and n-Engel groups.

DEFINITION 13.4. *We say that G is an* Engel group *if $G = R(G)$ or, equivalently, $G = L(G)$. Moreover, G is an n-*Engel *group if there exists $n \geq 1$ such that $[x, {}_n y] = 1$, for all $x, y \in G$.*

Of course every n-Engel group is Engel, and every nilpotent group of class n is n-Engel. Hence, the Engel condition is more general than nilpotency. Also, there are nilpotent groups of class n that are not $(n-1)$-Engel. For example, consider the groups $G = C_p \wr C_p$ (the wreath product of two cyclic groups of order a prime p). One can see that G is nilpotent of class p but not a $(p-1)$-Engel group (see Theorem 6.2 of [**146**]). Furthermore is not difficult to construct n-Engel groups that are not nilpotent. Take for example a prime p and consider the group $G = C_p \wr C_{p^\infty}$, where C_{p^∞} is the Prüfer group. One can prove that G is a $(p+1)$-Engel group but it is not nilpotent.

Recall that a group G is locally nilpotent if every finitely generated subgroup of G is nilpotent. The main open question in the realm of Engel groups is whether every n-Engel group is locally nilpotent.

For $n \leq 4$ every n-Engel group is locally nilpotent. Indeed, 1-Engel groups are abelian. For $n = 2$, Levi proved that G is a 2-Engel group if and only if the normal closure of an arbitrary element is abelian [**143**]. Moreover 2-Engel groups are nilpotent of class at most 3. For $n = 3$, Heineken proved that every 3-Engel group is locally nilpotent [**103**]. Finally, Havas and Vaughan-Lee [**101**] proved that every 4-Engel groups generated by d elements are nilpotent of class $4d$. For $n > 4$ the question is still open.

In some other classes of groups being n-Engel also implies local nilpotency. For example, any residually finite n-Engel group is locally nilpotent [**208**].

As pointed out in the introduction, the question whether every Engel group is locally nilpotent can be considered to be the analogue of the general Burnside problem for Engel groups.

It is easy to check that every locally nilpotent group is Engel. The converse is true in some classes of groups such as finite groups (Zorn [**213**]), soluble groups (Gruenberg [**93**]), groups with maximal condition (Baer [**6**]), linear groups ([**202**]) and some other classes of groups. Hence, in these cases the condition to being Engel is equivalent to local nilpotency, but in general it is much weaker. Indeed, the infinite finitely generated p-group constructed by Golod is a counterexample. This group is Engel and it is such that every $(r-1)$-generated subgroup is nilpotent. So far, in the periodic case, Golod's group is the only known example of finitely generated Engel group that is not nilpotent. See [**53**] for a good survey on Golod-Shafarevich groups.

In Chapter 2 we collect a list of algorithmic problems in groups. We refer the reader to the introduction of [**159**] for an exhaustive survey about algorithmic problems in nilpotent groups.

13.2.1. Identities satisfied by n-Engel groups. In this section we present some semigroup identities satisfied by n-Engel groups for $n \leq 4$. We have the following.

(a) In every 2-Engel group $yx^2y = xy^2x$.

(b) In every 3-Engel group

$$xy^2xyx^2y = yx^2yxy^2x \text{ and } xy^2xyxyx^2y = yx^2y^2x^2y^2x.$$

(c) In every 4-Engel group we have

$$xy^2xyx^2y^2x^2yxy^2xyx^2yxy^2x^2y^2xyx^2y = yx^2yxy^2x^2y^2xyx^2yxy^2xyx^2y^2x^2yxy^2x.$$

For more information one can see [**203**] and [**148**].

13.2.2. Burnside groups. Now we describe Burnside groups, since their behavior is similar to the behavior of the Engel groups. Moreover, in Section 13.3.3 we present a protocol that uses Burnside groups of exponent 3 as a platform group.

Let F_m be a free group of rank m. The free Burnside group $B(m, n)$ is the group F_m/F_m^n, where F_m^n is the group generated by all the n-th powers of elements of F_m. Therefore $B(m, n)$ is the group in which the identity $x^n = 1$ holds and for this reason is the *biggest* group generated by m elements of exponent n. The Burnside problem asks if a finitely generated group of finite exponent is finite, and this is equivalent to ask whether the free Burnside groups are finite.

It is easy to prove that for any m the 2-group $B(m, 2)$ is elementary abelian and hence finite. Burnside also proved that $B(m, 3)$ is finite for any m. In 1993 Levi and van der Waerden proved that if $m \geq 3$ the Burnside group $B(m, 3)$ is finite and nilpotent of class 3. One can also prove that since $B(m, 3)$ is a group of exponent 3, then it is a 2-Engel group because of the following.

PROPOSITION 13.5 (12.3.5 of [**186**]). A group of exponent 3 is a 2-Engel group.

In 1940, Sanov also proved that the group $B(m, 4)$ is finite for every m. For $n = 5$ the problem is still open. Although for some small values of n the Burnside groups are finite, Novikov and Adian proved in 1968 that, in general, Burnside groups need not be finite. Indeed they showed that if $m \geq 2$, and n is odd and greater than 4381, then $B(n, m)$ is infinite. This bound was improved later by Adian who showed that n can be chosen odd and greater than or equal to 665. With the same construction used to prove that these Burnside groups are infinite, Novikov and Adian also established that the word and the conjugacy problems are decidable in Burnside groups, see [**171**].

13.3. Cryptographic protocols

In this section we present some applications of Engel groups to cryptography.

13.3.1. Multilinear maps. In this subsection we provide a possible application of multilinear maps (see Chapter 3) in the context of Engel groups. For more information, see [**122**].

13.3.1.1. *Protocol.* Let G be a nilpotent group of class $n+1$ which is not n-Engel ($n \geq 1$). Then there exist elements $x, g \in G$ such that $[x,_n g] \neq 1$. The group G is public. We have $n+1$ users $A_1, \dots A_{n+1}$ that wish to agree on a shared secret key. Each user A_j selects a private integer $a_j \neq 0$, computes g^{a_j}, and sends it to the other users. Then we are in the following situation:

- The user A_1 computes $[x^{a_1}, g^{a_2}, \dots g^{a_{n+1}}]$.
- For $j=2,\dots,n$, the user A_j computes $[x^{a_j}, g^{a_1}, \dots, g^{a_{j-1}}, g^{a_{j+1}}, \dots, g^{a_{n+1}}]$.
- The user A_{n+1} computes $[x^{a_{n+1}}, g^{a_1}, \dots g^{a_n}]$.

Since (3.1) holds in all nilpotent groups, all elements computed by the users are equal to $k = [x,_n g]^{\prod_{j=1}^{n+1} a_j}$. This is the shared key.

13.3.2. Two cryptosystems based on n-th root problem in n-Engel groups. In this section we present two cryptosystems based on n-Engel groups proposed in [**203**].

13.3.2.1. *A public key based on 2-Engel groups.* A satellite generates some data from a 2-Engel group. Alice and Bob choose two elements x and y as their secret keys, respectively. Then Alice sends x^2 to Bob and Bob sends y^2 to Alice. As pointed out in Section 13.2.1, in any 2-Engel group the identity $xy^2x = yx^2y$ holds. Then Alice and Bob can use this as a shared key.

13.3.2.2. *A digital signature based on 4-Engel groups.* Consider a 4-Engel group, which is nilpotent and satisfies the following identity (see the last part of Section 13.2 and Section 13.2.1):

$$xy^2xyx^2y^2x^2yxy^2xyx^2yxy^2x^2y^2xyx^2y = yx^2yxy^2x^2y^2xyx^2yxy^2xyx^2y^2x^2yxy^2x.$$

The idea to make a digital signature is as follows. Suppose x and y are private and x^2, y^2, xy^2x, and $xy^2x^2y^2x$ are public. The public key is x^2 and the signature is xy^2x and $xy^2x^2y^2x$. The verifier knows y, and he has to check the identity above.

13.3.2.3. *Platform group and security.* The underlying group could be combined with a finite group in which finding square roots is hard, for example $\mathbb{Z}^*_{pq}$, with p, q prime numbers.

The security of this digital signatures relies on the complexity of finding square roots in 2-Engel and 4-Engel groups. It seems that this is a hard problem (see [**168**]).

13.3.3. Burnside groups of exponent 3 and the learning homomorphism problem. In [**56**] the authors used properties of Burnside groups in cryptography for the first time. In particular, they introduced the Learning *Burnside* Homomorphisms with Noise (B-LHN), which uses for computational purposes only surjective homomorphisms in contrast to the general definition of LHN [**12**] (see Chapter 2).

As a platform group for the B-LHN problem, the authors used the Burnside groups of exponent 3 that are, in particular, 2-Engel groups. The security of this protocol is based on the computational hardness of B-LHN (see Theorem 2 of [**56**]).

13.4. Interesting problems

As we have seen, there are a variety of key exchanges, digital signature systems, and secret sharing schemes for which an Engel group is an appropriate choice of platform group. As we have seen, on the one hand there has been some results about Engel groups and their proposed cryptosystems over the last decade. On the other hand the majority of computational complexity and algorithmic questions remain unanswered.

We collect some open problems below with the hope of stimulating interest in their solutions.

We start by pointing out that since Engel groups are a generalization of nilpotent groups, in the case in which the algorithmic problems presented are decidable for nilpotent groups, we ask whether the same holds specifically if the group is n-Engel. Also, since some problems are undecidable for nilpotent groups (take, for example, the rational subgroup membership problem and the endomorphism problem) we ask:

(a) Is the rational subgroup membership problem undecidable for Engel groups?
(b) Is the endomorphism problem undecidable for Engel groups?

In addition, we propose the following questions also related to the complexity of some algorithmic problems presented in Chapter 2.

(c) Determine if the following algorithmic problems are decidable and if yes, find the complexity:
    - the word problem,
    - the (power) conjugacy problem,
    - the geodesic length problem.
(d) What about the multiple conjugacy search problem? For example, if G is an Engel group, and $a_1, \dots a_n, b_1, \dots b_n \in G$ where a_i is conjugate to b_i for $i = 1, \dots, n$. Can we find $c \in G$ such that $a_i^c = b_i$ (for $1 \leq i \leq n$)? As a starting point, one can study the single conjugacy search problem (i.e., the case $n = 1$).

In Section 13.3 we present several ideas of Engel group-based cryptography.

In particular, in Section 13.3.1, we presented a protocol based on multilinear maps. The ideal platform of this scheme remains an open question. Thus we ask the following.

(e) What is an ideal platform group for the protocol proposed in Section 13.3.1?

In Chapter 13.3.2 we present some specific cryptosystems based on n-Engel groups. It seems they are secure because they are based on the difficulty to find square roots in n-Engel groups. Moreover, no platform groups are suggested. Whence, we ask the following.

(f) What is an ideal platform group for the protocol proposed in Section 13.3.2.1 and Section 13.3.2.2, respectively?
(g) What is the complexity of the square root problem in n-Engel groups? And for the n-root problem?
(h) Is the HSP solvable for Engel groups?
(i) Are there some Engel group-based cryptosystems resistant to quantum algorithms? Note that this could be done either by analyzing the HSP for Engel groups or showing the underlying security problem is NP-complete or NP-hard.

CHAPTER 14

Self-Similar Groups

14.1. Introduction

In this last chapter we present groups acting on regular rooted trees and, more precisely, the family of self-similar groups. These groups have been studied for years as an important source of groups with interesting properties. For instance, the Grigorchuk group (that is a self-similar group acting on the binary tree) was the first example of a finitely generated group with intermediate growth — this answered an open question posed by Milnor — and the first example of an amenable but not elementary amenable group. Furthermore, this group provides a counterexample to the general Burnside problem.

The chapter is organized as follows: in Section 14.2 we introduce the basic theory of groups of automorphisms of rooted trees and their subgroups. In Section 14.2.3 we present examples and main properties of such groups, including the aforementioned Grigorchuk group, the Grigorchuk-Gupta-Sidki (GGS) groups, and the Basilica group. Finally, we survey algorithmic problems in self-similar groups and the current status of self-similar groups in cryptography. We refer to [**9**] for a standard reference of self-similar groups.

14.2. Mathematical background

We start our exposition by describing the fundamental notion of a rooted tree.

14.2.1. Rooted trees. A tree is a connected graph with no cycles. A *rooted* tree is a tree with a designated vertex called the root. It is called *regular* if every vertex has the same valency (except for the root). If that valency is $d+1$, where d is an integer greater than or equal to 2, then d is the number of descendants of every vertex of the tree and we say that d is the *degree* of the tree. For every positive integer d there is a unique regular rooted tree of degree d that we will call simply the *d-adic tree*; see Figure 14.1.

We denote this tree with $\mathcal{T}_d$. When the value of d is clear from the context, we will simply write $\mathcal{T}$. Denote $X = \{x_1, x_2, \ldots, x_d\}$; then a *word* of length n in X as an expression of the form $x_{i_1}x_{i_2}\ldots x_{i_n}$, where $i_1, i_2, \ldots, i_n \in \{1, \ldots, d\}$. Let X^n be the set of all words of length n and $X^* = \bigcup_{n\in\mathbb{N}} X^n$ the set of all vertices of the tree. The free monoid generated by X is X^*, where we can multiply together elements by juxtaposition. Indeed, if we consider $u \in X^n$ and $v \in X^m$, then $uv \in X^{n+m}$. Moreover, chosen a vertex u, the subtree of $\mathcal{T}$ hanging from u constitutes the set uX^* and it is a copy of $\mathcal{T}$. It is customary to choose the set $X = \{0, 1, \ldots, d-1\}$.

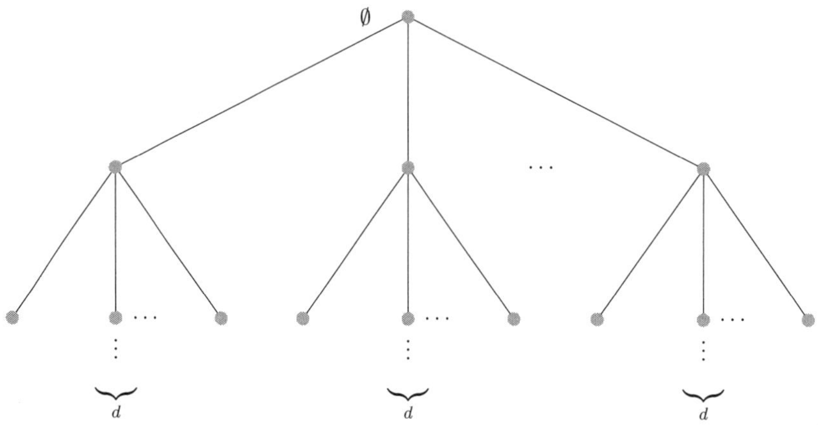

FIGURE 14.1

14.2.2. The group Aut $\mathcal{T}$ and some subgroups. If we consider bijective maps of the set X^* which preserve incidence, i.e., automorphisms of $\mathcal{T}$, the set of all of these maps is a group with respect to the ordinary composition between functions, where $(fg)(u) = g(f(u))$. We call it Aut $\mathcal{T}$. For example we have the identity map or we can permute rigidly the subtrees hanging from the vertices $0, 1, \ldots, d-1$ using a permutation $\alpha \in \mathrm{Sym}(d)$, where $\mathrm{Sym}(d)$ denotes the symmetric group on d letters. This is called the *rooted automorphism* corresponding to α. Let g be in Aut $\mathcal{T}$. If g sends u to v then

$$\begin{aligned} u0 &\longmapsto v\alpha(0), \\ u1 &\longmapsto v\alpha(1), \\ &\vdots \\ ud-1 &\longmapsto v\alpha(d-1), \end{aligned}$$

where α is a permutation. Furthermore, $\alpha \in \mathrm{Sym}(d)$ is called the *label* of g at the vertex u and it is denoted with $g_{(u)}$. The *portrait* of g is the set of all labels of g, and gives a way of describing automorphisms. Indeed, given $g \in$ Aut $\mathcal{T}$, if its portrait is known, then we can calculate the image under g of every vertex of $\mathcal{T}$. More precisely,

$$g(i_1 i_2 \ldots i_n) = g_{(\emptyset)}(i_1) g_{(i_1)}(i_2) \ldots g_{(i_1 \ldots i_{n-1})}(i_n). \tag{14.1}$$

Conversely, given any portrait on the tree with elements in $\mathrm{Sym}(d)$, by formula (14.1) we can define an automorphism of $\mathcal{T}$.

Now we are going to define two important subgroups of Aut $\mathcal{T}$. The *n-th level stabilizer* of Aut $\mathcal{T}$ is

$$\mathrm{St}(n) = \{f \in \mathrm{Aut}\,\mathcal{T} \mid f(u) = u \ \forall u \in X^n\}.$$

More generally, if $H \leq$ Aut $\mathcal{T}$, we define $\mathrm{St}_H(n) = H \cap \mathrm{St}(n)$.

Let $f \in$ Aut $\mathcal{T}$ and $u \in X^*$ and suppose that $f(u) = u$. The *section* of f at u is the restriction of f to the subtree hanging from u, which is identified with $\mathcal{T}$. It is denoted by f_u, and can be described as $f(uv) = uf_u(v)$. Using sections, one can introduce the following map, which will be very important in the remainder of the text.

If $n \in \mathbb{N}$, there is an isomorphism

$$\psi_n : \mathrm{St}(n) \longrightarrow \mathrm{Aut}\,\mathcal{T} \times \overset{d^n}{\cdots} \times \mathrm{Aut}\,\mathcal{T}$$
$$g \longmapsto (g_u)_{u\in X^n},$$

for every $g \in \mathrm{St}(n)$.

For this reason we have:

$$\mathrm{St}(n) \simeq \mathrm{Aut}\,\mathcal{T} \times \overset{d^n}{\cdots} \times \mathrm{Aut}\,\mathcal{T}.$$

This result implies that $\mathrm{Aut}\,\mathcal{T}$ contains direct products $\mathrm{Aut}\,\mathcal{T} \times \overset{d^n}{\cdots} \times \mathrm{Aut}\,\mathcal{T}$, a fact that is not true for a general group (if we consider $\mathbb{Z}$, there does not exist a subgroup H of $\mathbb{Z}$ such that $H \simeq \mathbb{Z} \times \mathbb{Z}$).

Observe that by giving the image of $\psi_n(g)$, we can use ψ_n to define an element $g \in \mathrm{St}(n)$. We usually write ψ instead of ψ_1.

Notice that the semidirect structure of $\mathrm{Aut}\,\mathcal{T}$ and the fact that $\mathrm{St}(n) \cong \mathrm{Aut}\,\mathcal{T} \times \overset{d^n}{\cdots} \times \mathrm{Aut}\,\mathcal{T}$ imply that $\mathrm{Aut}\,\mathcal{T}$ can also be seen as the iterated permutational wreath product:

$$\mathrm{Aut}\,\mathcal{T} \cong \mathrm{Sym}(d) \ltimes \mathrm{St}(1) \cong \mathrm{Aut}\,\mathcal{T} \wr \mathrm{Sym}(d) \cong ((\cdots \wr \mathrm{Sym}(d)) \wr \mathrm{Sym}(d)) \wr \mathrm{Sym}(d).$$

From this it follows that an element $g \in \mathrm{Aut}\,\mathcal{T}$ can be written in the form $g = h\sigma$, where $h = (h_1, \ldots, h_d) \in \mathrm{St}(1)$, and $\sigma \in \mathrm{Sym}(d)$.

We conclude this section by defining the groups of interest in this chapter.

DEFINITION 14.1. *A group* $G \leq \mathrm{Aut}\,\mathcal{T}$ *is* self-similar *if* $\psi(h) \in G \times \overset{d}{\cdots} \times G$ *for all* $g = h\sigma \in G$.

For example, $\mathrm{Aut}\mathcal{T}$ is self-similar.

14.2.3. Examples of self-similar groups. In this section we provide some examples of well-known self-similar groups, such as the (first) Grigorchuk group, the GGS groups, and the the Basilica group. For ease of readability we make the abuse of notation of writing an element omitting the map ψ. For example, we write $a = (1, b)$ instead of $\psi(a) = (1, b)$.

The first Grigorchuk group. The (first) Grigorchuk group $\mathfrak{G}$, introduced by Grigorchuk [**83**] in 1980, is a subgroup of $\mathrm{Aut}\,\mathcal{T}_2$. The Grigorchuk group possesses a lot of interesting properties. For instance, it was the first group shown to be of intermediate growth [**84**]; it is a counterexample to the general Burnside problem; and it is amenable but not elementary amenable [**85**]. For more information about growth of groups and amenability see [**155**] and [**176**], respectively.

The group is generated by $a = (1\ 2)$ that is the swap at the root, and three other elements $b, c, d \in \mathrm{St}(1)$ defined recursively as follows

$$b = (a, c), \quad c = (a, d), \quad d = (1, b),$$

whose portrait is represented in Figure 14.2.

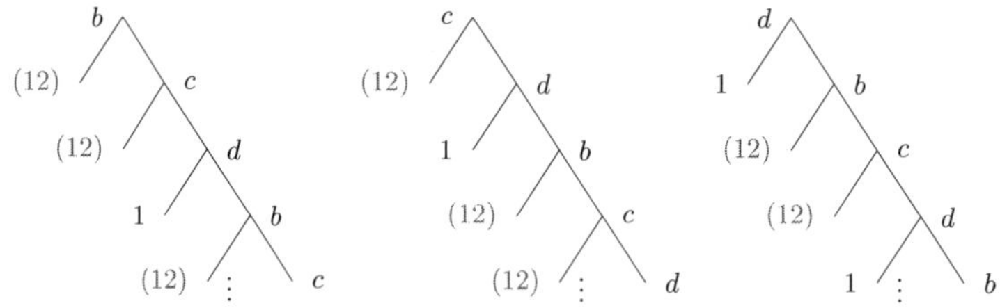

FIGURE 14.2. Portrait of the elements b, c, and d of the Grigorchuk group.

For further information and for the proof of the following theorem, one can see [**38**, Chapter 7].

THEOREM 14.2. *The group* $\mathfrak{G}$ *satisfies the following properties:*
(i) $\mathfrak{G}$ *is self-similar.*
(ii) $\mathfrak{G}$ *is amenable, but not elementary amenable.*
(iii) $\mathfrak{G}$ *is a 2-group.*
(iv) $\mathfrak{G}$ *is a group of intermediate growth.*

The Grigorchuk groups. The family of Grigorchuk groups G_ω acts on the binary tree $\mathcal{T}_2$ and is defined via the parameter $\omega = (\omega_1, \omega_2, \dots)$, with $\omega_i \in \{0, 1, 2\}$ for any $i \in \mathbb{N}$. We denote Ω as the set of all such ω and for $n \in \mathbb{N}$, we write $\sigma^n\omega$ for the sequence $(\omega_{n+1}, \omega_{n+2}, \dots)$.

For $\omega \in \Omega$, the Grigorchuk group G_ω is generated by $a, b_\omega, c_\omega, d_\omega$, where a is the rooted automorphism corresponding to the cycle $(1\,2)$, and the directed automorphisms $b_\omega, c_\omega, d_\omega$ lie in the first level stabilizer and are defined as follows. For $n \in \mathbb{N} \cup \{0\}$, let $u_n = 1 \overset{n}{\dots} 1 \in X^*$. Then $b_\omega(u_n 0)$ is trivial if $\omega_{n+1} = 2$ and is equal to a otherwise; the automorphism $c_\omega(u_n 0)$ is trivial if $\omega_{n+1} = 1$ and is equal to a otherwise; and finally $d_\omega(u_n 0)$ is trivial if $\omega_{n+1} = 0$ and is equal to a otherwise. More precisely:

$$\varphi_u(b_\omega) = \begin{cases} a & \text{if } u = u_n 0 \text{ for some } n \text{ and } \omega_{n+1} \neq 2, \\ e & \text{if } u = u_n 0 \text{ for some } n \text{ and } \omega_{n+1} = 2, \\ b_{\sigma^n \omega} & \text{if } u = u_n \text{ for some } n; \end{cases}$$

$$\varphi_u(c_\omega) = \begin{cases} a & \text{if } u = u_n 0 \text{ for some } n \text{ and } \omega_{n+1} \neq 1, \\ e & \text{if } u = u_n 0 \text{ for some } n \text{ and } \omega_{n+1} = 1, \\ c_{\sigma^n \omega} & \text{if } u = u_n \text{ for some } n; \end{cases}$$

$$\varphi_u(d_\omega) = \begin{cases} a & \text{if } u = u_n 0 \text{ for some } n \text{ and } \omega_{n+1} \neq 0, \\ e & \text{if } u = u_n 0 \text{ for some } n \text{ and } \omega_{n+1} = 0, \\ d_{\sigma^n \omega} & \text{if } u = u_n \text{ for some } n. \end{cases}$$

For example, if we take the sequence $(0, \omega_2, \omega_3, \dots)$, we obtain

$$b_\omega = (a, b_{\sigma\omega}), \quad c_\omega = (a, c_{\sigma\omega}), \quad \text{and} \quad d_\omega = (e, d_{\sigma\omega}).$$

If $\omega = \{0, 1, 2, 0, 1, 2, \dots\}$, then G_ω is the first Grigorchuk group defined in the previous section.

The GGS groups. Let p be an odd prime, and let $\mathbf{e} = (e_1, \ldots, e_{p-1})$ be a vector, where $e_i \in \{1, \ldots, p\}$ and such that not all the e_i's are 0. The GGS group $\mathcal{G}_{\mathbf{e}}$, named after Grigorchuk, Gupta, and Sidki, is the group generated by the two automorphisms $a, b \in \operatorname{Aut} \mathcal{T}_p$, where a is the rooted automorphism corresponding to the cycle $(1 \ldots p)$, and $b \in \operatorname{St}(1)$ is

$$b = (a^{e_1}, \ldots, a^{e_{p-1}}, b).$$

For example the famous Gupta-Sidki p-group is a GGS group whose defining vector $\mathbf{e}$ is $\mathbf{e} = (1, -1, 0, \ldots, 0)$. The group $\mathcal{G}_{\mathbf{e}}$ is self-similar for any choice of the vector $\mathbf{e}$, but other properties depend on the choice of $\mathbf{e}$ (see [**57**]).

The Basilica group. The Basilica group $\mathcal{B}$ was introduced by Grigorchuk and Zuk in 2002 [**88**], and it is the subgroup of $\operatorname{Aut} \mathcal{T}_2$ generated by the two automorphisms a and b defined as

$$a = (1, b) \quad \text{and} \quad b = (1, a)\sigma,$$

where σ is the rooted automorphism of $\mathcal{T}_2$ that corresponds to the permutation $(1\ 2)$. Here, we abuse notation by writing $a = (1, b)$ instead of $\psi(a) = (1, b)$, and similarly for the element $(1, a)$. The portrait of these two automorphisms is represented in Figure 14.3.

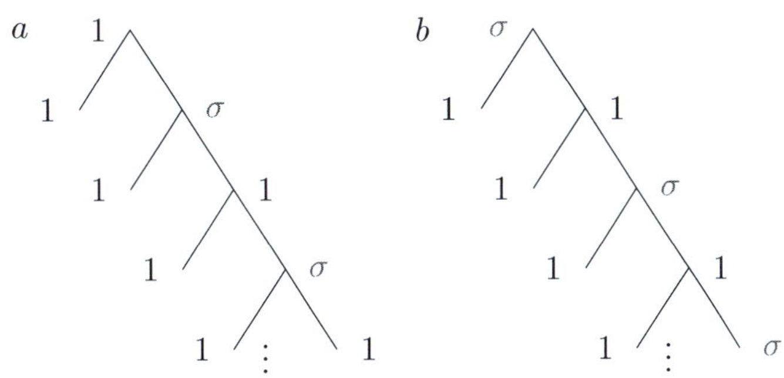

FIGURE 14.3. The portrait of the generators of the Basilica group $\mathcal{B}$.

Recently, this group has been generalized to the wider class of p-Basilica groups [**42**] generated by the two automorphisms a and b defined as

$$a = (1, \ldots, 1, b) \quad \text{and} \quad b = (1, \ldots, 1, a)\sigma.$$

Notice that the 2-Basilica group is the one defined before, and it is straightforward to see that all these groups are self-similar.

14.3. Algorithmic problems

In 1984 Grigorchuk proved that the word problem is solvable in some groups of automorphisms of rooted trees called spinal groups by using a fast universal algorithm. In particular, all contracting branch groups (one of the most studied class of self-similar groups) have solvable word problem. Although most branch groups have been shown to have decidable word problem, there exists some spinal group that has undecidable word problem [**85**]. In 1997 Wilson and Zaleskii solved the conjugacy problem in GGS p-groups (using the notion of Malcev's conjugacy separability and pro-p methods) for p odd [**209**], while in 2000 Grigorchuk and Wilson proved that the conjugacy problem is solvable for a wide class of branch groups [**87**]. Leonov [**141**] and Rozhkov [**189**] independently solved the case $p = 2$.

The standard approach to algorithmic problems in groups considers finitely presented groups. Bartholdi [**8**] in 2003 proved that contracting branch groups are not finitely presented but, in spite of this, they admit an endomorphic presentation (i.e., a recursive presentation) that allows a more general approach to the study of algorithmic problems.

Concerning complexity, it has been determined that the conjugacy problem in the Grigorchuk group has polynomial time complexity. In 2017 it has been proved that the conjugacy problem in the Grigorchuk group has log-space complexity [**165**].

14.4. Cryptographic protocols

The first protocol based on Grigorchuk groups was introduced in [**71**], although it was shown to be insecure in [**178**]. In a recent paper [**86**], Grigorchuk and Grigoriev suggest some branch groups as platforms for Anshel-Anshel-Goldfeld key-agreement. Among others, they suggest some branch groups like the aforementioned Grigorchuk group and the Basilica group.

In the following we give some details on the aforementioned protocols built around self-similar groups that we mentioned above.

14.4.1. Public-key cryptosystem based on the word problem. In [**71**] the authors proposed a public-key cryptosystem based on the word problem in the family of Grigorchuk groups. This protocol is similar to one constructed by Wagner and Magyarik [**207**], and it is the first application of groups acting on rooted trees in cryptography. The protocol works as follows.

Let G_ω be a Grigorchuk group, where $\omega \in \{0,1,2\}^{\mathbb{N}}$.

- Alice chooses her private element $\omega \in \{0,1,2\}^{\mathbb{N}}$. Alice's public key is a finite set of words that denote the identity element of G_ω, and two different words a, b.
- Bob encrypts the message by choosing a bit $i \in \{0,1\}$ and then he obtains a word w_i^* by randomly applying a series of additions or deletions of relations provided by Alice.
- Alice decrypts the message by checking whether $w_i = w_0$ or $w_i = 1$.

In [**71**] the authors declared that this public-key cryptosystem is secure since there is not enough information to determine the key ω.

Attack. The cryptosystem presented above was first attacked by Hofheinz and Steinwandt in [**108**]. However their algorithm is mainly based on brute force. Later in 2003, Petrides [**178**] claimed that this protocol is insecure since the public key given gives too much information and allows a possible attacker to easily obtain the private key.

14.4.2. Key agreement scheme based on the simultaneous conjugacy problem. In [**86**], the authors proposed the Anshel-Anshel-Goldfeld protocol, whose security is based on the difficulty of the subgroup-restricted simultaneous conjugacy problem. The protocol is as follows.

Let G be a group. and consider public elements $a_1, \ldots, a_n, b_1, \ldots, b_m \in G$. The Anshel–Anshel–Goldfeld key agreement works as follows:

- Alice chooses her private element $a = a_{p_1} \ldots a_{p_s} \in \langle a_1, \ldots, a_n \rangle$.
- Bob chooses his private element $b = b_{q_1} \ldots b_{q_t} \in \langle b_1, \ldots, b_m \rangle$.
- Alice sends $b_i^a = a^{-1}ba$, with $1 \leq i \leq m$.
- Bob sends $a_j^b = b^{-1}a_j b$, with $1 \leq j \leq n$.
- Alice computes $bab^{-1} = ba_{p_1}b^{-1} \ldots ba_{p_s}b^{-1}$
- Bob computes $a^{-1}ba = a^{-1}b_{q_1}a^{-1} \ldots a^{-1}b_{q_t}a$.

The common key is the commutator. Finally, the commutator $[a, b^{-1}] = a^{-1}(bab^{-1}) = (a^{-1}ba)b^{-1}$ computed by Alice and Bob.

An adversary has to find elements $A \in \langle a_1, \ldots, a_n \rangle$ and $B \in \langle b_1, \ldots, b_m \rangle$ such that $A^{-1}b_i A = a^{-1}b_i a$ with $1 \leq i \leq m$ and $Ba_j B^{-1} = ba_j b^{-1}$ with $1 \leq j \leq n$, where $a^{-1}b_i a$ and $ba_j b^{-1}$ are known. Finally, one has to check that $a^{-1}bab^{-1} = A^{-1}BAB^{-1}$. Furthermore, a possible attacker has to search a solution $A \in \langle a_1, \ldots, a_n \rangle$ of the problem $A^{-1}b_i A = a^{-1}b_i a$ with $1 \leq i \leq m$. The ideal platform groups for this protocol are those in which the complexity of solving the simultaneous conjugacy problem is hard, but with polynomial time complexity of the word problem. As an example, one can take the aforementioned p-Basilica groups.

Bibliography

[1] G. Alagic and A. Russell, *Quantum-secure symmetric-key cryptography based on hidden shifts*, Advances in cryptology—EUROCRYPT 2017. Part III, Lecture Notes in Comput. Sci., vol. 10212, Springer, Cham, 2017, pp. 65–93, DOI 10.1007/978-3-319-56617-7_3. MR3652147

[2] I. Anshel, M. Anshel, and D. Goldfeld, *An algebraic method for public-key cryptography*, Math. Res. Lett. **6** (1999), no. 3-4, 287–291, DOI 10.4310/MRL.1999.v6.n3.a3. MR1713130

[3] I. Anshel, D. Atkins, D. Goldfeld, and P. E. Gunnells, *WalnutDSA™: a group theoretic digital signature algorithm*, Int. J. Comput. Math. Comput. Syst. Theory **6** (2021), no. 4, 260–284, DOI 10.1080/23799927.2020.1831613. MR4355498

[4] Y. Antolín and A. W. Sale, *Permute and conjugate: the conjugacy problem in relatively hyperbolic groups*, Bull. Lond. Math. Soc. **48** (2016), no. 4, 657–675, DOI 10.1112/blms/bdw028. MR3532141

[5] L. Babai, *Group, graphs, algorithms: the graph isomorphism problem*, Proceedings of the International Congress of Mathematicians—Rio de Janeiro 2018. Vol. IV. Invited lectures, World Sci. Publ., Hackensack, NJ, 2018, pp. 3319–3336. MR3966534

[6] R. Baer, *Engelsche elemente Noetherscher Gruppen*, Math. Ann. **133** (1957), 256–270, https://doi.org/10.1007/BF02547953.

[7] J. Barnard, N. Brady, and P. Dani, *Super-exponential distortion of subgroups of $CAT(-1)$ groups*, Algebr. Geom. Topol. **7** (2007), 301–308, DOI 10.2140/agt.2007.7.301. MR2308946

[8] L. Bartholdi, *Endomorphic presentations of branch groups*, J. Algebra **268** (2003), no. 2, 419–443, DOI 10.1016/S0021-8693(03)00268-0. MR2009317

[9] L. Bartholdi, R. I. Grigorchuk, and Z. Šunić, *Branch groups*, Handbook of algebra, Vol. 3, Handb. Algebr., vol. 3, Elsevier/North-Holland, Amsterdam, 2003, pp. 989–1112, DOI 10.1016/S1570-7954(03)80078-5. MR2035113

[10] C. Battarbee, D. Kahrobaei, and S. F. Shahandashti, *Cryptanalysis of semidirect product key exchange using matrices over non-commutative rings*, Mathematical Cryptology **1** (2021), no. 1, 1–8.

[11] C. Battarbee, D. Kahrobaei, L. Perret, and S. F Shahandashti, *A subexponential quantum algorithm for the semdirect discrete logarithm problem*, arXiv preprint arXiv:2209.02814 (2022).

[12] G. Baumslag, N. Fazio, A. R. Nicolosi, V. Shpilrain, and W. E. Skeith III, *Generalized learning problems and applications to non-commutative cryptography (extended abstract)*, Provable security, Lecture Notes in Comput. Sci., vol. 6980, Springer, Heidelberg, 2011, pp. 324–339, DOI 10.1007/978-3-642-24316-5_23. MR2911480

[13] W. Beullens and S. R. Blackburn, *Practical attacks against the walnut digital signature scheme*, Advances in cryptology—ASIACRYPT 2018. Part I, Lecture Notes in Comput. Sci., vol. 11272, Springer, Cham, 2018, pp. 35–61, DOI 10.1007/978-3-030-03326-2_2. MR3897822

[14] J. S. Birman, *Braids, links, and mapping class groups*, Annals of Mathematics Studies, No. 82, Princeton University Press, Princeton, NJ; University of Tokyo Press, Tokyo, 1974. MR375281

[15] J. S. Birman, K. H. Ko, and S. J. Lee, *A new approach to the word and conjugacy problems in the braid groups*, Adv. Math. **139** (1998), no. 2, 322–353, https://doi.org/10.1006/aima.1998.1761. MR1654165

[16] D. Boneh and R. J. Lipton, *Quantum cryptanalysis of hidden linear functions (extended abstract)*, Advances in cryptology—CRYPTO '95 (Santa Barbara, CA, 1995), Lecture Notes in Comput. Sci., vol. 963, Springer, Berlin, 1995, pp. 424–437, DOI 10.1007/3-540-44750-4_34. MR1445578

[17] D. Boneh and A. Silverberg, *Applications of multilinear forms to cryptography*, Topics in algebraic and noncommutative geometry (Luminy/Annapolis, MD, 2001), 2003, pp. 71–90, `https://doi.org/10.1090/conm/324/05731`.

[18] A. Borel, *Introduction aux groupes arithmétiques* (French), Publications de l'Institut de Mathématique de l'Université de Strasbourg, XV. Actualités Scientifiques et Industrielles, No. 1341, Hermann, Paris, 1969. MR244260

[19] M. R. Bridson, *On the subgroups of right-angled Artin groups and mapping class groups*, Math. Res. Lett. **20** (2013), no. 2, 203–212, DOI 10.4310/MRL.2013.v20.n2.a1. MR3151642

[20] M. R. Bridson and A. Haefliger, *Metric spaces of non-positive curvature*, Grundlehren der mathematischen Wissenschaften [Fundamental Principles of Mathematical Sciences], vol. 319, Springer-Verlag, Berlin, 1999, DOI 10.1007/978-3-662-12494-9. MR1744486

[21] M. R. Bridson and J. Howie, *Conjugacy of finite subsets in hyperbolic groups*, Internat. J. Algebra Comput. **15** (2005), no. 4, 725–756, DOI 10.1142/S0218196705002529. MR2160576

[22] D. R. L. Brown, N. Koblitz, and J. T. LeGrow, *Cryptanalysis of "MAKE"*, J. Math. Cryptol. **16** (2022), no. 1, 98–102, DOI 10.1515/jmc-2021-0016. MR4378544

[23] S. Burckel, *The well-ordering on positive braids*, J. Pure Appl. Algebra **120** (1997), no. 1, 1–17, DOI 10.1016/S0022-4049(96)00072-2. MR1466094

[24] W. Burnside, *On an unsettled question in the theory of discontinous groups*, Quart. J. Pure Appl. Math. **33** (1902), 230–238.

[25] M. Casals-Ruiz, *Embeddability and quasi-isometric classification of partially commutative groups*, Algebr. Geom. Topol. **16** (2016), no. 1, 597–620, DOI 10.2140/agt.2016.16.597. MR3470711

[26] W. Castryck, T. Lange, C. Martindale, L. Panny, and J. Renes, *CSIDH: an efficient post-quantum commutative group action*, Advances in cryptology—ASIACRYPT 2018. Part III, Lecture Notes in Comput. Sci., vol. 11274, Springer, Cham, 2018, pp. 395–427, DOI 10.1007/978-3-030-03332-3_15. MR3897883

[27] B. Cavallo and D. Kahrobaei, *Secret sharing using non-commutative groups and the shortlex order*, Algorithmic problems of group theory, their complexity, and applications to cryptography, Contemp. Math., vol. 633, Amer. Math. Soc., Providence, RI, 2015, pp. 1–8, DOI 10.1090/conm/633/12646. MR3364217

[28] B. Chandler and W. Magnus, *The history of combinatorial group theory: A case study in the history of ideas*, Studies in the History of Mathematics and Physical Sciences, vol. 9, Springer-Verlag, New York, 1982, DOI 10.1007/978-1-4613-9487-7. MR680777

[29] R. Charney, *An introduction to right-angled Artin groups*, Geom. Dedicata **125** (2007), 141–158, DOI 10.1007/s10711-007-9148-6. MR2322545

[30] I. Chatterji, D. Kahrobaei, and N. Y. Lu, *Cryptosystems using subgroup distortion*, Theoretical and Applied Informatics **29** (2017), no. 1-2, 14–24.

[31] J. H. Cheon and B. Jun, *A polynomial time algorithm for the braid Diffie-Hellman conjugacy problem*, Advances in cryptology—CRYPTO 2003, Lecture Notes in Comput. Sci., vol. 2729, Springer, Berlin, 2003, pp. 212–225, DOI 10.1007/978-3-540-45146-4_13. MR2093195

[32] A. Childs, *Lecture notes on quantum algorithms, https://www.cs.umd.edu/ amchilds/qa/qa.pdf* (2017), `http://www.cs.umd.edu/~amchilds/qa/qa.pdf`.

[33] A. M. Childs and W. van Dam, *Quantum algorithms for algebraic problems*, Rev. Modern Phys. **82** (2010), no. 1, 1–52, `https://doi.org/10.1103/RevModPhys.82.1`. MR2629607

[34] A. M. Childs and G. Ivanyos, *Quantum computation of discrete logarithms in semigroups*, J. Math. Cryptol. **8** (2014), no. 4, 405–416, DOI 10.1515/jmc-2013-0038. MR3281663

[35] A. M. Childs and W. van Dam, *Quantum algorithms for algebraic problems*, Rev. Modern Phys. **82** (2010), no. 1, 1–52, DOI 10.1103/RevModPhys.82.1. MR2629607

[36] J.-M. Couveignes, *Hard homogeneous spaces*, Cryptology ePrint Archive (2006).

[37] J. Crisp, E. Godelle, and B. Wiest, *The conjugacy problem in subgroups of right-angled Artin groups*, J. Topol. **2** (2009), no. 3, 442–460, DOI 10.1112/jtopol/jtp018. MR2546582

[38] P. de la Harpe, *Topics in geometric group theory*, Chicago Lectures in Mathematics, University of Chicago Press, Chicago, IL, 2000. MR1786869

[39] P. Dehornoy, *A fast method for comparing braids*, Adv. Math. **125** (1997), no. 2, 200–235, DOI 10.1006/aima.1997.1605. MR1434111

[40] P. Dehornoy, *Braid-based cryptography*, Group theory, statistics, and cryptography, Contemp. Math., vol. 360, Amer. Math. Soc., Providence, RI, 2004, pp. 5–33, DOI 10.1090/conm/360/06566. MR2105432

[41] P. Dehornoy, I. Dynnikov, D. Rolfsen, and B. Wiest, *Ordering braids*, Mathematical Surveys and Monographs, vol. 148, American Mathematical Society, Providence, RI, 2008, DOI 10.1090/surv/148. MR2463428

[42] E. Di Domenico, G. A. Fernández-Alcober, M. Noce, and A. Thillaisundaram, *p-Basilica groups*, Mediterr. J. Math. **19** (2022), no. 6, Paper No. 275, 28, DOI 10.1007/s00009-022-02187-z. MR4505206

[43] W. Diffie and M. E. Hellman, *New directions in cryptography*, IEEE Trans. Inform. Theory **IT-22** (1976), no. 6, 644–654, https://doi.org/10.1109/tit.1976.1055638.

[44] B. Domanski and M. Anshel, *The complexity of Dehn's algorithm for word problems in groups*, J. Algorithms **6** (1985), no. 4, 543–549, DOI 10.1016/0196-6774(85)90031-8. MR813591

[45] C. Droms, *Isomorphisms of graph groups*, Proc. Amer. Math. Soc. **100** (1987), no. 3, 407–408, https://doi.org/10.2307/2046419.

[46] B. Eick and D. Kahrobaei, *Polycyclic groups: A new platform for cryptology?*, arXiv math:0411077 (2004).

[47] B. Eick and G. Ostheimer, *On the orbit-stabilizer problem for integral matrix actions of polycyclic groups*, Math. Comp. **72** (2003), no. 243, 1511–1529, DOI 10.1090/S0025-5718-03-01493-5. MR1972750

[48] K. Eisenträger, S. Hallgren, A. Kitaev, and F. Song, *A quantum algorithm for computing the unit group of an arbitrary degree number field*, STOC'14—Proceedings of the 2014 ACM Symposium on Theory of Computing, ACM, New York, 2014, pp. 293–302. MR3238955

[49] E. A. El-Rifai and H. R. Morton, *Algorithms for positive braids*, Quart. J. Math. Oxford Ser. (2) **45** (1994), no. 180, 479–497, https://doi.org/10.1093/qmath/45.4.479.

[50] T. ElGamal, *A public key cryptosystem and a signature scheme based on discrete logarithms*, IEEE Trans. Inform. Theory **31** (1985), no. 4, 469–472, DOI 10.1109/TIT.1985.1057074. MR798552

[51] D. B. A. Epstein, J. W. Cannon, D. F. Holt, S. V. F. Levy, M. S. Paterson, and W. P. Thurston, *Word processing in groups*, Jones and Bartlett Publishers, Boston, MA, 1992. MR1161694

[52] D. B. A. Epstein and D. F. Holt, *The linearity of the conjugacy problem in word-hyperbolic groups*, Internat. J. Algebra Comput. **16** (2006), no. 2, 287–305, https://doi.org/10.1142/S0218196706002986. MR2228514

[53] M. Ershov, *Golod-Shafarevich groups: a survey*, Internat. J. Algebra Comput. **22** (2012), no. 5, 1230001, 68, DOI 10.1142/S0218196712300010. MR2949205

[54] B. Farb, *The extrinsic geometry of subgroups and the generalized word problem*, Proc. London Math. Soc. (3) **68** (1994), no. 3, 577–593, https://doi.org/10.1112/plms/s3-68.3.577. MR1262309

[55] D. Farley and L. Sabalka, *Discrete Morse theory and graph braid groups*, Algebr. Geom. Topol. **5** (2005), 1075–1109, https://doi.org/10.2140/agt.2005.5.1075. MR2171804

[56] N. Fazio, K. Iga, A. R. Nicolosi, L. Perret, and W. E. Skeith III, *Hardness of learning problems over Burnside groups of exponent 3*, Des. Codes Cryptogr. **75** (2015), no. 1, 59–70, https://doi.org/10.1007/s10623-013-9892-6. MR3320351

[57] G. A. Fernández-Alcober and A. Zugadi-Reizabal, *GGS-groups: order of congruence quotients and Hausdorff dimension*, Trans. Amer. Math. Soc. **366** (2014), no. 4, 1993–2017, DOI 10.1090/S0002-9947-2013-05908-9. MR3152720

[58] B. Fine, M. Habeeb, D. Kahrobaei, and G. Rosenberger, *Aspects of nonabelian group-based cryptography: a survey and open problems*, JP J. Algebra Number Theory Appl. **21** (2011), no. 1, 1–40. MR2840945

[59] B. Fine, A. Moldenhauer, and G. Rosenberger, *Cryptographic protocols based on Nielsen transformations*, Journal of Computer and Communication **4** (2016), 63–107.

[60] R. Flores and D. Kahrobaei, *Cryptography with right-angled Artin groups*, Theoretical and Applied Informatics **28** (2016), no. 3, 8–16.

[61] R. Flores, D. Kahrobaei, and T. Koberda, *Algorithmic problems in right-angled Artin groups: complexity and applications*, J. Algebra **519** (2019), 111–129, DOI 10.1016/j.jalgebra.2018.10.023. MR3874519

[62] R. Flores, D. Kahrobaei, and T. Koberda, *A cryptographic application of the Thurston norm*, Int. J. Comput. Math. Comput. Syst. Theory **5** (2020), no. 1, 15–24, DOI 10.1080/23799927.2020.1716074. MR4095613

[63] R. Flores, D. Kahrobaei, and T. Koberda, *Expander graphs and right-angled Artin groups*, arXiv 2005:06143 (2022).

[64] E. Formanek, *Conjugate separability in polycyclic groups*, J. Algebra **42** (1976), no. 1, 1–10, DOI 10.1016/0021-8693(76)90021-1. MR419608

[65] N. Franco and J. González-Meneses, *Conjugacy problem for braid groups and Garside groups*, J. Algebra **266** (2003), no. 1, 112–132, DOI 10.1016/S0021-8693(03)00292-8. MR1994532

[66] K. Friedl, G. Ivanyos, F. Magniez, M. Santha, and P. Sen, *Hidden translation and translating coset in quantum computing*, SIAM J. Comput. **43** (2014), no. 1, 1–24, https://doi.org/10.1137/130907203.

[67] D. Garber, D. Kahrobaei, and H. T. Lam, *Length-based attacks in polycyclic groups*, J. Math. Cryptol. **9** (2015), no. 1, 33–43, DOI 10.1515/jmc-2014-0003. MR3318546

[68] D. Garber, S. Kaplan, M. Teicher, B. Tsaban, and U. Vishne, *Length-based conjugacy search in the braid group*, Algebraic methods in cryptography, 2006, pp. 75–87, https://doi.org/10.1090/conm/418/07947.

[69] P. B. Garrett, *Making, breaking codes: Introduction to cryptology*, 1st ed., Prentice Hall PTR, USA, 2000.

[70] F. A. Garside, *The braid group and other groups*, Quart. J. Math. Oxford Ser. (2) **20** (1969), 235–254, DOI 10.1093/qmath/20.1.235. MR248801

[71] M. Garzon and Y. Zalcstein, *The complexity of Grigorchuk groups with application to cryptography*, Theoret. Comput. Sci. **88** (1991), no. 1, 83–98, DOI 10.1016/0304-3975(91)90074-C. MR1130373

[72] V. Gebhardt, *A new approach to the conjugacy problem in Garside groups*, J. Algebra **292** (2005), no. 1, 282–302, DOI 10.1016/j.jalgebra.2005.02.002. MR2166805

[73] V. Gebhardt and J. González-Meneses, *Solving the conjugacy problem in Garside groups by cyclic sliding*, J. Symbolic Comput. **45** (2010), no. 6, 629–656, https://doi.org/10.1016/j.jsc.2010.01.013.

[74] S. M. Gersten and H. B. Short, *Rational subgroups of biautomatic groups*, Ann. of Math. (2) **134** (1991), no. 1, 125–158, DOI 10.2307/2944334. MR1114609

[75] É. Ghys and P. de la Harpe (eds.), *Sur les groupes hyperboliques d'après Mikhael Gromov* (French), Papers from the Swiss Seminar on Hyperbolic Groups held in Bern, 1988, Progress in Mathematics, vol. 83, Birkhäuser Boston, Inc., Boston, MA, 1990, DOI 10.1007/978-1-4684-9167-8. MR1086648

[76] R. Gitik, *Nielsen generating sets and quasiconvexity of subgroups*, J. Pure Appl. Algebra **112** (1996), no. 3, 287–292, https://doi.org/10.1016/0022-4049(96)00148-X. MR1410179

[77] S. Gogioso and A. Kissinger, *Fully graphical treatment of the quantum algorithm for the hidden subgroup problem*, arXiv 1701:08669 (2017).

[78] J. González-Meneses, *Improving an algorithm to solve multiple simultaneous conjugacy problems in braid groups*, Geometric methods in group theory, Contemp. Math., vol. 372, Amer. Math. Soc., Providence, RI, 2005, pp. 35–42, DOI 10.1090/conm/372/06872. MR2139675

[79] J. González-Meneses, *Basic results on braid groups*, Ann. Math. Blaise Pascal **18** (2011), no. 1, 15–59, https://doi.org/10.5802/ambp.293.

[80] J. González-Meneses and M. Silvero, *Polynomial braid combing*, Math. Comp. **88** (2019), no. 318, 2027–2045, https://doi.org/10.1090/mcom/3392.

[81] E. Green, *Graph products of groups*, PhD Thesis (1990).

[82] E. Greendlinger and M. Greendlinger, *On Dehn presentations and Dehn algorithms*, Illinois J. Math. **30** (1986), no. 2, 360–363. MR840134

[83] R. I. Grigorčuk, *On Burnside's problem on periodic groups* (Russian), Funktsional. Anal. i Prilozhen. **14** (1980), no. 1, 53–54. MR565099

[84] R. I. Grigorchuk, *On the Milnor problem of group growth* (Russian), Dokl. Akad. Nauk SSSR **271** (1983), no. 1, 30–33. MR712546

[85] R. I. Grigorchuk, *Degrees of growth of finitely generated groups and the theory of invariant means* (Russian), Izv. Akad. Nauk SSSR Ser. Mat. **48** (1984), no. 5, 939–985. MR764305

[86] R. Grigorchuk and D. Grigoriev, *Key agreement based on automaton groups*, Groups Complex. Cryptol. **11** (2019), no. 2, 77–81, DOI 10.1515/gcc-2019-2012. MR4026474

[87] R. I. Grigorchuk and J. S. Wilson, *The conjugacy problem for certain branch groups* (English, with Russian summary), Tr. Mat. Inst. Steklova **231** (2000), no. Din. Sist., Avtom. i Beskon. Gruppy, 215–230; English transl., Proc. Steklov Inst. Math. **4(231)** (2000), 204–219. MR1841756

[88] R. I. Grigorchuk and A. Żuk, *On a torsion-free weakly branch group defined by a three state automaton*, Internat. J. Algebra Comput. **12** (2002), no. 1-2, 223–246, DOI 10.1142/S0218196702001000. International Conference on Geometric and Combinatorial Methods in Group Theory and Semigroup Theory (Lincoln, NE, 2000). MR1902367

[89] D. Grigoriev, *Testing shift-equivalence of polynomials by deterministic, probabilistic and quantum machines*, Theoret. Comput. Sci. **180** (1997), no. 1-2, 217–228, DOI 10.1016/S0304-3975(96)00188-0. MR1453867

[90] D. Grigoriev and I. Ponomarenko, *Constructions in public-key cryptography over matrix groups*, Algebraic methods in cryptography, Contemp. Math., vol. 418, Amer. Math. Soc., Providence, RI, 2006, pp. 103–119, DOI 10.1090/conm/418/07949. MR2389292

[91] D. Grigoriev and V. Shpilrain, *Authentication schemes from actions on graphs, groups, or rings*, Ann. Pure Appl. Logic **162** (2010), no. 3, 194–200, DOI 10.1016/j.apal.2010.09.004. MR2740680

[92] M. Gromov, *Hyperbolic groups*, Essays in group theory, 1987, pp. 75–263, https://doi.org/10.1007/978-1-4613-9586-7_3.

[93] K. W. Gruenberg, *Two theorems on Engel groups*, Proc. Cambridge Philos. Soc. **49** (1953), 377–380, https://doi.org/10.1017/s0305004100028528.

[94] J. Gryak and D. Kahrobaei, *The status of polycyclic group-based cryptography: a survey and open problems*, Groups Complex. Cryptol. **8** (2016), no. 2, 171–186, DOI 10.1515/gcc-2016-0013. MR3567873

[95] W. H. Gustafson, *What is the probability that two group elements commute?*, Amer. Math. Monthly **80** (1973), 1031–1034, DOI 10.2307/2318778. MR327901

[96] M. Habeeb, D. Kahrobaei, C. Koupparis, and V. Shpilrain, *Public key exchange using semidirect product of (semi) groups*, Applied cryptography and network security, 2013, pp. 475–486.

[97] M. Habeeb, D. Kahrobaei, and V. Shpilrain, *A secret sharing scheme based on group presentations and the word problem*, Computational and combinatorial group theory and cryptography, Contemp. Math., vol. 582, Amer. Math. Soc., Providence, RI, 2012, pp. 143–150, DOI 10.1090/conm/582/11557. MR2987392

[98] F. Haglund and D. T. Wise, *Special cube complexes*, Geom. Funct. Anal. **17** (2008), no. 5, 1551–1620, https://doi.org/10.1007/s00039-007-0629-4.

[99] D. Hart, D. Kim, G. Micheli, G. Pascual-Perez, C. Petit, and Y. Quek, *A practical cryptanalysis of WalnutDSA*^{TM}, Public-key cryptography—PKC 2018. Part I, Lecture Notes in Comput. Sci., vol. 10769, Springer, Cham, 2018, pp. 381–406, DOI 10.1007/978-3-319-76578-5_13. MR3781016

[100] K. Hauschild and W. Rautenberg, *Interpretierbarkeit in der Gruppentheorie*, Algebra Universalis **1** (1971/72), 136–151, https://doi.org/10.1007/BF02944970.

[101] G. Havas and M. R. Vaughan-Lee, *4-Engel groups are locally nilpotent*, Internat. J. Algebra Comput. **15** (2005), no. 4, 649–682, DOI 10.1142/S0218196705002475. MR2160572

[102] H. Heineken, *Eine Bemerkung über engelsche Elemente* (German), Arch. Math. (Basel) **11** (1960), 321, DOI 10.1007/BF01236951. MR125153

[103] H. Heineken, *Engelsche Elemente der Länge drei*, Illinois J. Math. **5** (1961), 681–707, http://projecteuclid.org/euclid.ijm/1255631589. MR0131469

[104] S. Hermiller and J. Meier, *Algorithms and geometry for graph products of groups*, J. Algebra **171** (1995), no. 1, 230–257, https://doi.org/10.1006/jabr.1995.1010. MR1314099

[105] A. Herzberg, M. Jakobsson, S. Jarecki, H. Krawczyk, and M. Yung, *Proactive public key and signature systems*, Proceedings of the 4th ACM conference on Computer and communications security, CCS '97 (1997), 100–110.

[106] M. Hirvensalo, *Quantum computing*, 2nd ed., Natural Computing Series, Springer-Verlag, Berlin, 2004, DOI 10.1007/978-3-662-09636-9. MR2248224

[107] J. Hoffstein, J. Pipher, and J. H. Silverman, *An introduction to mathematical cryptography*, 2nd ed., Undergraduate Texts in Mathematics, Springer, New York, 2014, DOI 10.1007/978-1-4939-1711-2. MR3289167

[108] D. Hofheinz and R. Steinwandt, *Cryptanalysis of a public key cryptosystem based on Grigorchuk group*, Unpublished.

[109] D. Hofheinz and R. Steinwandt, *A practical attack on some braid group-based cryptographic primitives*, Public key cryptography—PKC 2003, Lecture Notes in Comput. Sci., vol. 2567, Springer, Berlin, 2002, pp. 187–198, DOI 10.1007/3-540-36288-6_14. MR2171926

[110] D. F. Holt, B. Eick, and E. A. O'Brien, *Handbook of computational group theory*, Discrete Mathematics and its Applications (Boca Raton), Chapman & Hall/CRC, Boca Raton, FL, 2005, DOI 10.1201/9781420035216. MR2129747

[111] J. M. Howie, *Fundamentals of semigroup theory*, London Mathematical Society Monographs. New Series, vol. 12, The Clarendon Press, Oxford University Press, New York, 1995. Oxford Science Publications. MR1455373

[112] J. Hughes, *The left SSS attack on Ko-Lee-Cheon-Han-Kang-Park key agreement protocol in B45*, Rump Session, Crypto 2000 (2000).

[113] J. Hughes and A. Tannenbaum, *Length-based attacks for certain group based encryption rewriting systems, Workshop SECI02 Sécurité de la communication sur internet*, 2002.

[114] G. Ivanyos, F. Magniez, and M. Santha, *Efficient quantum algorithms for some instances of the non-abelian hidden subgroup problem*, 2003, pp. 723–739, https://doi.org/10.1142/S0129054103001996.

[115] D. Kahrobaei and B. Khan, *Nis05-6: A non-commutative generalization of ElGamal key exchange using polycyclic groups*, IEEE Globecom, 2006, pp. 1–5.

[116] D. Kahrobaei and C. Koupparis, *Non-commutative digital signatures*, Groups Complex. Cryptol. **4** (2012), no. 2, 377–384, DOI 10.1515/gcc-2012-0019. MR3043439

[117] D. Kahrobaei and H. T. Lam, *Heisenberg groups as platform for the AAG key-exchange protocol*, 22nd International Conference on Network Protocols (ICNP), IEEE (2014), 660–664.

[118] D. Kahrobaei and K. Mallahi-Karai, *Some applications of arithmetic groups in cryptography*, Groups Complex. Cryptol. **11** (2019), no. 1, 25–33, DOI 10.1515/gcc-2019-2002. MR4000595

[119] D. Kahrobaei and M. Noce, *Algorithmic problems in Engel groups and cryptographic applications*, Int. J. Group Theory **9** (2020), no. 4, 231–250. MR4223607

[120] D. Kahrobaei and V. Shpilrain, *Using semidirect product of (semi)groups in public key cryptography*, Computability in europe 2016: Pursuit of the universal, 2016, pp. 132–141, https://doi.org/10.1007/978-3-319-40189-8_14.

[121] D. Kahrobaei and M. Stanojkovski, *Cryptographic multilinear maps using pro-p groups*, Adv. Math. Commun. **17** (2023), no. 5, 1101–1114, DOI 10.3934/amc.2021041. MR4576170

[122] D. Kahrobaei, A. Tortora, and M. Tota, *Multilinear cryptography using nilpotent groups*, Elementary theory of groups and group rings, and related topics, De Gruyter Proc. Math., De Gruyter, Berlin, [2020] ©2020, pp. 127–133, DOI 10.1515/9783110638387-013. MR4204987

[123] I. Kapovich, *Detecting quasiconvexity: algorithmic aspects*, Geometric and computational perspectives on infinite groups (Minneapolis, MN and New Brunswick, NJ, 1994), DIMACS Ser. Discrete Math. Theoret. Comput. Sci., vol. 25, Amer. Math. Soc., Providence, RI, 1996, pp. 91–99, DOI 10.1090/dimacs/025/07. MR1364182

[124] I. Kapovich and N. Benakli, *Boundaries of hyperbolic groups*, Combinatorial and geometric group theory (New York, 2000/Hoboken, NJ, 2001), 2002, pp. 39–93, https://doi.org/10.1090/conm/296/05068.

[125] I. Kapovich, A. Miasnikov, P. Schupp, and V. Shpilrain, *Average-case complexity and decision problems in group theory*, Adv. Math. **190** (2005), no. 2, 343–359, DOI 10.1016/j.aim.2003.02.001. MR2102661

[126] C. Kassel and V. Turaev, *Braid groups*, Graduate Texts in Mathematics, vol. 247, Springer, New York, 2008, https://doi.org/10.1007/978-0-387-68548-9.

[127] A. Keromytis and M. Yung (eds.), *Applied cryptography and network security*, ACNS 2005, Lecture Notes in Comput. Sci., vol. 3531, Springer, Berlin, Heidelberg.

[128] S. Kijima, Y. Otachi, T. Saitoh, and T. Uno, *Subgraph isomorphism in graph classes*, Discrete Math. **312** (2012), no. 21, 3164–3173, DOI 10.1016/j.disc.2012.07.010. MR2957936

[129] S.-h. Kim and T. Koberda, *Embedability between right-angled Artin groups*, Geom. Topol. **17** (2013), no. 1, 493–530, DOI 10.2140/gt.2013.17.493. MR3039768

[130] A. Kitaev, *Quantum computations: algorithms and error correction*, Uspekhi Mat. Nauk **52** (1997), no. 6(318), 53–112, https://doi.org/10.1070/RM1997v052n06ABEH002155.

[131] K. Ko, S. Lee, J. Cheon, J. Han, J. Kang, and C. Park, *New public-key cryptosystem using braid groups*, Advances in cryptology—CRYPTO 2000 (Santa Barbara, CA), 2000, pp. 166–183, `https://doi.org/10.1007/3-540-44598-6_10`.

[132] K. H. Ko, J. W. Lee, and T. Thomas, *Towards generating secure keys for braid cryptography*, Des. Codes Cryptogr. **45** (2007), no. 3, 317–333, DOI 10.1007/s10623-007-9123-0. MR2350533

[133] T. Koberda, *Right-angled groups and their subgroups*, Unpublished notes, https://faculty.virginia.edu/Koberda/raagcourse.pdf.

[134] J. Köbler, U. Schöning, and J. Torán, *The graph isomorphism problem: its structural complexity*, Progress in Theoretical Computer Science, Birkhäuser Boston, Inc., Boston, MA, 1993, DOI 10.1007/978-1-4612-0333-9. MR1232421

[135] N. Koblitz, *Algebraic aspects of cryptography*, with an appendix by Alfred J. Menezes, Yi-Hong Wu, and Robert J. Zuccherato, Algorithms Comput. Math. **3** (1998), Springer-Verlag, Berlin. MR1610535

[136] M. Kotov, A. Menshov, and A. Ushakov, *An attack on the Walnut digital signature algorithm*, Des. Codes Cryptogr. **87** (2019), no. 10, 2231–2250, `https://doi.org/10.1007/s10623-019-00615-y`.

[137] M. Kotov and A. Ushakov, *Analysis of a certain polycyclic-group-based cryptosystem*, J. Math. Cryptol. **9** (2015), no. 3, 161–167, DOI 10.1515/jmc-2015-0013. MR3403498

[138] K. Kulesza and Z. Kotulski, *Addressing new challenges by building security protocols around graphs*, Security protocols, Lecture Notes in Comput. Sci., vol. 3364, Springer, Berlin, 2005, pp. 301–306, DOI 10.1007/11542322_36. MR2214280

[139] G. Kuperberg, *A subexponential-time quantum algorithm for the dihedral hidden subgroup problem*, SIAM J. Comput. **35** (2005), no. 1, 170–188, DOI 10.1137/S0097539703436345. MR2178804

[140] S. J. Lee and E. Lee, *Potential weaknesses of the commutator key agreement protocol based on braid groups*, Advances in cryptology—EUROCRYPT 2002 (Amsterdam), Lecture Notes in Comput. Sci., vol. 2332, Springer, Berlin, 2002, pp. 14–28, DOI 10.1007/3-540-46035-7_2. MR1975525

[141] Y. G. Leonov, *The conjugacy problem in a class of* 2*-groups*, Mat. Zametki **64** (1998), no. 4, 573–583, `https://doi.org/10.1007/BF02314631`.

[142] P. Lescot, *Isoclinism classes and commutativity degrees of finite groups*, J. Algebra **177** (1995), no. 3, 847–869, `https://doi.org/10.1006/jabr.1995.1331`.

[143] F. W. Levi, *Groups in which the commutator operation satisfies certain algebraic conditions*, J. Indian Math. Soc. (N.S.) **6** (1942), 87–97. MR7417

[144] L. Levin, *Universal sorting problems*, Problems of Information Transmission **9** (1973), 265–266.

[145] F. Levy-dit Vehel and L. Perret, *On the Wagner-Magyarik cryptosystem*, Coding and cryptography, 2006, pp. 316–329, `https://doi.org/10.1007/11779360_25`.

[146] H. Liebeck, *Concerning nilpotent wreath products*, Proc. Cambridge Philos. Soc. **58** (1962), 443–451. MR139656

[147] H.-N. Liu, C. Wrathall, and K. Zeger, *Efficient solution of some problems in free partially commutative monoids*, Inform. and Comput. **89** (1990), no. 2, 180–198, DOI 10.1016/0890-5401(90)90010-F. MR1080848

[148] P. Longobardi and M. Maj, *Semigroup identities and Engel groups*, Groups St. Andrews 1997 in Bath, II, London Math. Soc. Lecture Note Ser., vol. 261, Cambridge Univ. Press, Cambridge, 1999, pp. 527–531, DOI 10.1017/CBO9780511666148.016. MR1676649

[149] A. Lubotzky, S. Mozes, and M. S. Raghunathan, *The word and Riemannian metrics on lattices of semisimple groups*, Inst. Hautes Études Sci. Publ. Math. **91** (2000), 5–53 (2001). MR1828742

[150] R. C. Lyndon and P. E. Schupp, *Combinatorial group theory*, Classics in Mathematics, Springer-Verlag, Berlin, 2001. Reprint of the 1977 edition, DOI 10.1007/978-3-642-61896-3. MR1812024

[151] K. Madlener and F. Otto, *Using string-rewriting for solving the word problem for finitely presented groups*, Inform. Process. Lett. **24** (1987), no. 5, 281–284, DOI 10.1016/0020-0190(87)90149-9. MR882625

[152] A. Mal'cev, *On homomorphisms onto finite groups*, Trans. Amer. Math. Soc **119** (1983), 67–79.

[153] T. Mecham and A. Mukherjee, *Hyperbolic groups which fiber in infinitely many ways*, Algebr. Geom. Topol. **9** (2009), no. 4, 2101–2120, DOI 10.2140/agt.2009.9.2101. MR2551664

[154] J. Milnor, *Growth of finitely generated solvable groups*, J. Differential Geometry **2** (1968), 447–449, `http://projecteuclid.org/euclid.jdg/1214428659`.

[155] J. Milnor, *A note on curvature and fundamental group*, J. Differential Geometry **2** (1968), 1–7, `http://projecteuclid.org/euclid.jdg/1214501132`.

[156] M. R. R. Moghaddam, A. R. Salemkar, and K. Chiti, *n-isoclinism classes and n-nilpotency degree of finite groups*, Algebra Colloq. **12** (2005), no. 2, 255–261, DOI 10.1142/S1005386705000246. MR2127250

[157] G Moussong, *Hyperbolic Coxeter groups*, ProQuest LLC, Ann Arbor, MI, 1988. Thesis (Ph.D.)–The Ohio State University, `http://gateway.proquest.com/openurl?url_ver=Z39.88-2004&rft_val_fmt=info:ofi/fmt:kev:mtx:dissertation&res_dat=xri:pqdiss&rft_dat=xri:pqdiss:8824577`.

[158] A. Myasnikov, V. Shpilrain, and A. Ushakov, *A practical attack on a braid group based cryptographic protocol*, Advances in cryptology—CRYPTO 2005, Lecture Notes in Comput. Sci., vol. 3621, Springer, Berlin, 2005, pp. 86–96, DOI 10.1007/11535218_6. MR2237301

[159] A. Myasnikov and A. Weiß, TC^0 *circuits for algorithmic problems in nilpotent groups*, 42nd International Symposium on Mathematical Foundations of Computer Science, LIPIcs. Leibniz Int. Proc. Inform., vol. 83, Schloss Dagstuhl. Leibniz-Zent. Inform., Wadern, 2017, pp. Art. No. 23, 14. MR3755316

[160] A. Myasnikov, V. Shpilrain, and A. Ushakov, *Non-commutative cryptography and complexity of group-theoretic problems*, with an appendix by Natalia Mosina, Mathematical Surveys and Monographs, vol. 177, American Mathematical Society, Providence, RI, 2011, DOI 10.1090/surv/177. MR2850384

[161] A. Myasnikov and V. Roman'kov, *A linear decomposition attack*, Groups Complex. Cryptol. **7** (2015), no. 1, 81–94, `https://doi.org/10.1515/gcc-2015-0007`.

[162] A. Myasnikov, V. Shpilrain, and A. Ushakov, *Non-commutative cryptography and complexity of group-theoretic problems*, with an appendix by Natalia Mosina, Mathematical Surveys and Monographs, vol. 177, American Mathematical Society, Providence, RI, 2011, DOI 10.1090/surv/177. MR2850384

[163] A. Myasnikov and A. Ushakov, *Length based attack and braid groups: cryptanalysis of Anshel-Anshel-Goldfeld key exchange protocol*, Public key cryptography—PKC 2007, Lecture Notes in Comput. Sci., vol. 4450, Springer, Berlin, 2007, pp. 76–88, DOI 10.1007/978-3-540-71677-8_6. MR2404113

[164] A. Myasnikov and A. Ushakov, *Random subgroups and analysis of the length-based and quotient attacks*, J. Math. Cryptol. **2** (2008), no. 1, 29–61, DOI 10.1515/JMC.2008.003. MR2451658

[165] A. Miasnikov and S. Vassileva, *Log-space conjugacy problem in the Grigorchuk group*, Groups Complex. Cryptol. **9** (2017), no. 1, 77–85, DOI 10.1515/gcc-2017-0005. MR3641682

[166] L. Narendra-Mohan and G. S. G. N. Anjaneyulu, *A secured digital signature using conjugacy and DLP on non-commutative group over finite field*, Advances in Intelligent Systems and Computing (AISC) **516** (2017), 449–465.

[167] B. B. Newman, *Some results on one-relator groups*, Bull. Amer. Math. Soc. **74** (1968), 568–571, DOI 10.1090/S0002-9904-1968-12012-9. MR222152

[168] W. Nickel, *Computation of nilpotent Engel groups*, J. Austral. Math. Soc. Ser. A **67** (1999), no. 2, 214–222. Group theory. MR1717415

[169] M. A. Nielsen and I. L. Chuang, *Quantum computation and quantum information*, Cambridge University Press, Cambridge, 2000. MR1796805

[170] NIST, *Post-Quantum Cryptography PQC*, 2020. `https:\csrc.nist.gov/projects/post-quantum-cryptography`.

[171] P. S. Novikov and S. I. Adjan, *Infinite periodic groups. I–III*, Mathematics of the USSR-Izvestiya **2** (1968).

[172] E. O'Brien, *Polycyclic groups*. `https://www.icts.res.in/sites/default/files/1284624569notes.pdf`.

[173] R. O'Donnell, *Quantum computation and quantum information*, 2018, `https://www.cs.cmu.edu/~odonnell/quantum18/`.

[174] R. Ostrovsky and W. E. Skeith III, *Communication complexity in algebraic two-party protocols*, Advances in cryptology—CRYPTO 2008, Lecture Notes in Comput. Sci., vol. 5157, Springer, Berlin, 2008, pp. 379–396, DOI 10.1007/978-3-540-85174-5_21. MR2490385

[175] D. Panagopoulos, *A secret sharing scheme using groups*, arXiv 1009.0026v1 (2010).

[176] A. L. T. Paterson, *Amenability*, Mathematical Surveys and Monographs, vol. 29, American Mathematical Society, Providence, RI, 1988, DOI 10.1090/surv/029. MR961261

[177] C. Petit and J. J. Quisquater, *Cryptographic hash functions and expander graphs: the end of the story?*, The new codebreakers, 2016, pp. 304–311, `https://doi.org/10.1007/978-3-662-49301-4_19`.

[178] G. Petrides, *Cryptanalysis of the public key cryptosystem based on the word problem on the Grigorchuk groups*, Cryptography and coding, Lecture Notes in Comput. Sci., vol. 2898, Springer, Berlin, 2003, pp. 234–244, DOI 10.1007/978-3-540-40974-8_19. MR2090936

[179] S. J. Pride, *Small cancellation conditions satisfied by one-relator groups*, Math. Z. **184** (1983), no. 2, 283–286, DOI 10.1007/BF01252863. MR716277

[180] N. Rahman and V. Shpilrain, *MAKE: a matrix action key exchange*, J. Math. Cryptol. **16** (2022), no. 1, 64–72, DOI 10.1515/jmc-2020-0053. MR4360035

[181] O. Regev, *Quantum computation and lattice problems*, SIAM J. Comput. **33** (2004), no. 3, 738–760, `https://doi.org/10.1137/S0097539703440678`.

[182] O. Regev, *On lattices, learning with errors, random linear codes, and cryptography*, J. ACM **56** (2009), no. 6, Art. 34, 40, `https://doi.org/10.1145/1568318.1568324`.

[183] V. N. Remeslennikov, *Conjugacy in polycyclic groups* (Russian), Algebra i Logika **8** (1969), 712–725. MR280593

[184] E. Rips, *Subgroups of small cancellation groups*, Bull. London Math. Soc. **14** (1982), no. 1, 45–47, DOI 10.1112/blms/14.1.45. MR642423

[185] R. L. Rivest, A. Shamir, and L. Adleman, *A method for obtaining digital signatures and public-key cryptosystems*, Comm. ACM **21** (1978), no. 2, 120–126, `https://doi.org/10.1145/359340.359342`.

[186] D. J. S. Robinson, *A course in the theory of groups*, Second, Graduate Texts in Mathematics, vol. 80, Springer-Verlag, New York, 1996, `https://doi.org/10.1007/978-1-4419-8594-1`.

[187] M. Roetteler and T. Beth, *Polynomial-time solution to the hidden subgroup problem for a class of non-abelian groups*, arXiv quant-ph:9812070 (1998).

[188] V. Roman'kov, *A linear decomposition attack on public key exchange protocols using semidirect products of (semi)groups*, arXiv 1501:01152 (2015).

[189] A. V. Rozhkov, *The conjugacy problem in an automorphism group of an infinite tree*, Mat. Zametki **64** (1998), no. 4, 592–597, `https://doi.org/10.1007/BF02314633`.

[190] P. E. Schupp, *A survey of small cancellation theory*, Word problems: decision problems and the Burnside problem in group theory (Conf., Univ. California, Irvine, Calif., 1969; dedicated to Hanna Neumann), Stud. Logic Found. Math., Vol. 71, North-Holland, Amsterdam-London, 1973, pp. 569–589. MR412289

[191] D. Segal, *Decidable properties of polycyclic groups*, Proc. London Math. Soc. (3) **61** (1990), no. 3, 497–528, DOI 10.1112/plms/s3-61.3.497. MR1069513

[192] Z. Sela, *The isomorphism problem for hyperbolic groups. I*, Ann. of Math. (2) **141** (1995), no. 2, 217–283, `https://doi.org/10.2307/2118520`.

[193] H. Servatius, *Automorphisms of graph groups*, J. Algebra **126** (1989), no. 1, 34–60, `https://doi.org/10.1016/0021-8693(89)90319-0`.

[194] P. W. Shor, *Algorithms for quantum computation: discrete logarithms and factoring*, 35th Annual Symposium on Foundations of Computer Science (Santa Fe, NM, 1994), IEEE Comput. Soc. Press, Los Alamitos, CA, 1994, pp. 124–134, DOI 10.1109/SFCS.1994.365700. MR1489242

[195] V. Shpilrain and A. Ushakov, *The conjugacy search problem in public key cryptography: unnecessary and insufficient*, Appl. Algebra Engrg. Comm. Comput. **17** (2006), no. 3-4, 285–289, DOI 10.1007/s00200-006-0009-6. MR2233787

[196] V. Shpilrain and A. Ushakov, *A new key exchange protocol based on the decomposition problem*, Algebraic methods in cryptography, Contemp. Math., vol. 418, Amer. Math. Soc., Providence, RI, 2006, pp. 161–167, DOI 10.1090/conm/418/07954. MR2389297

[197] V. Shpilrain and G. Zapata, *Combinatorial group theory and public key cryptography*, Appl. Algebra Engrg. Comm. Comput. **17** (2006), no. 3-4, 291–302, DOI 10.1007/s00200-006-0006-9. MR2233788

[198] H. Sibert, P. Dehornoy, and M. Girault, *Entity authentication schemes using braid word reduction*, Discrete Appl. Math. **154** (2006), no. 2, 420–436, DOI 10.1016/j.dam.2005.03.015. MR2194414

[199] C. Staecker, *Dehn.gap, http://cstaecker.fairfield.edu/ cstaecker/files/gap/dehn.gap.*

[200] D. R. Stinson, *Cryptography*, 3rd ed., Discrete Mathematics and its Applications (Boca Raton), Chapman & Hall/CRC, Boca Raton, FL, 2006. Theory and practice. MR2182472

[201] R. Strebel, *Appendix. Small cancellation groups*, Sur les groupes hyperboliques d'après Mikhael Gromov (Bern, 1988), Progr. Math., vol. 83, Birkhäuser Boston, Boston, MA, 1990, pp. 227–273. MR1086661

[202] D. A. Suprunenko and M. S. Garaščuk, *Linear groups with Engel's condition* (Russian), Dokl. Akad. Nauk BSSR **6** (1962), 277–280. MR143818

[203] S. Sze, D. Kahrobaei, R. Dambreville, and M. Dupas, *Finding n-th roots in nilpotent groups and applications in cryptology*, Int. J. Pure Appl. Math. **70** (2011), no. 4, 571–593. MR2848462

[204] T. tom Dieck, *Transformation groups*, De Gruyter Studies in Mathematics, vol. 8, Walter de Gruyter & Co., Berlin, 1987, DOI 10.1515/9783110858372.312. MR889050

[205] B. Tsaban, *Polynomial-time solutions of computational problems in noncommutative-algebraic cryptography*, J. Cryptology **28** (2015), no. 3, 601–622, DOI 10.1007/s00145-013-9170-9. MR3355794

[206] L. Van Wyk, *Graph groups are biautomatic*, J. Pure Appl. Algebra **94** (1994), no. 3, 341–352, `https://doi.org/10.1016/0022-4049(94)90015-9`.

[207] N. R. Wagner and M. R. Magyarik, *A public-key cryptosystem based on the word problem*, Advances in cryptology (Santa Barbara, Calif., 1984), 1985, pp. 19–36, `https://doi.org/10.1007/3-540-39568-7_3`.

[208] J. S. Wilson, *Two-generator conditions for residually finite groups*, Bull. London Math. Soc. **23** (1991), no. 3, 239–248, `https://doi.org/10.1112/blms/23.3.239`.

[209] J. S. Wilson and P. A. Zalesskii, *Conjugacy separability of certain torsion groups*, Arch. Math. (Basel) **68** (1997), no. 6, 441–449, `https://doi.org/10.1007/s000130050076`.

[210] J. A. Wolf, *Growth of finitely generated solvable groups and curvature of Riemannian manifolds*, J. Differential Geometry **2** (1968), 421–446, `http://projecteuclid.org/euclid.jdg/1214428658`.

[211] C. Wrathall, *The word problem for free partially commutative groups*, J. Symbolic Comput. **6** (1988), no. 1, 99–104, DOI 10.1016/S0747-7171(88)80024-5. MR961373

[212] P. Zhu and Q. Wen, *Affine braid groups: a better platform than braid groups for cryptology?*, Appl. Algebra Engrg. Comm. Comput. **22** (2011), no. 5-6, 375–391, DOI 10.1007/s00200-011-0157-1. MR2863392

[213] M. Zorn, *Nilpotency of finite groups*, Bull. Amer. Math. Soc. **42** (1936), 485–486.

Index

SELECTED PUBLISHED TITLES IN THIS SERIES

278 **Delaram Kahrobaei, Ramón Flores, Marialaura Noce, Maggie E. Habeeb, and Christopher Battarbee,** Applications of Group Theory in Cryptography, 2024

277 **George Metcalfe, Francesco Paoli, and Constantine Tsinakis,** Residuated Structures in Algebra and Logic, 2023

276 **Balázs Bárány, Károly Simon, and Boris Solomyak,** Self-similar and Self-affine Sets and Measures, 2023

275 **Alekos Vidras and Alain Yger,** Multidimensional Residue Theory and Applications, 2023

274 **Willi Freeden and M. Zuhair Nashed,** Recovery Methodologies: Regularization and Sampling, 2023

273 **Gennadiy Feldman,** Characterization of Probability Distributions on Locally Compact Abelian Groups, 2023

272 **Tadashi Ochiai,** Iwasawa Theory and Its Perspective, Volume 1, 2023

271 **Hideto Asashiba,** Categories and Representation Theory, 2022

270 **Leslie Hogben, Jephian C.-H. Lin, and Bryan L. Shader,** Inverse Problems and Zero Forcing for Graphs, 2022

269 **Ralph S. Freese, Ralph N. McKenzie, George F. McNulty, and Walter F. Taylor,** Algebras, Lattices, Varieties, 2022

268 **Ralph S. Freese, Ralph N. McKenzie, George F. McNulty, and Walter F. Taylor,** Algebras, Lattices, Varieties, 2022

267 **Dragana S. Cvetković Ilić,** Completion Problems on Operator Matrices, 2022

266 **Kate Juschenko,** Amenability of Discrete Groups by Examples, 2022

265 **Nabil H. Mustafa,** Sampling in Combinatorial and Geometric Set Systems, 2022

264 **J. Scott Carter and Seiichi Kamada,** Diagrammatic Algebra, 2021

263 **Vugar E. Ismailov,** Ridge Functions and Applications in Neural Networks, 2021

262 **Ragnar-Olaf Buchweitz,** Maximal Cohen–Macaulay Modules and Tate Cohomology, 2021

261 **Shiri Artstein-Avidan, Apostolos Giannopoulos, and Vitali D. Milman,** Asymptotic Geometric Analysis, Part II, 2021

260 **Lindsay N. Childs, Cornelius Greither, Kevin P. Keating, Alan Koch, Timothy Kohl, Paul J. Truman, and Robert G. Underwood,** Hopf Algebras and Galois Module Theory, 2021

259 **William Heinzer, Christel Rotthaus, and Sylvia Wiegand,** Integral Domains Inside Noetherian Power Series Rings, 2021

258 **Pramod N. Achar,** Perverse Sheaves and Applications to Representation Theory, 2021

257 **Juha Kinnunen, Juha Lehrbäck, and Antti Vähäkangas,** Maximal Function Methods for Sobolev Spaces, 2021

256 **Michio Jimbo, Tetsuji Miwa, and Fedor Smirnov,** Local Operators in Integrable Models I, 2021

255 **Alexandre Boritchev and Sergei Kuksin,** One-Dimensional Turbulence and the Stochastic Burgers Equation, 2021

254 **Karim Belabas and Henri Cohen,** Numerical Algorithms for Number Theory, 2021

253 **Robert R. Bruner and John Rognes,** The Adams Spectral Sequence for Topological Modular Forms, 2021

252 **Julie Déserti,** The Cremona Group and Its Subgroups, 2021

251 **David Hoff,** Linear and Quasilinear Parabolic Systems, 2020

250 **Bachir Bekka and Pierre de la Harpe,** Unitary Representations of Groups, Duals, and Characters, 2020

For a complete list of titles in this series, visit the
AMS Bookstore at **www.ams.org/bookstore/survseries/**.